In large commercial buildings and industrial plants, the trend is to use 12,470 V to 24,000 V distribution systems. Power at this voltage is delivered to substations within the building where transformers step down the voltage to the level needed.

TRANSFORMERS

A *transformer* is a device which steps up or steps down alternating voltages. The current is varied in the opposite direction by the same proportion when the voltage in a circuit is stepped up or stepped down by a transformer. A transformer does not change one form of power into another like a generator or motor.

A transformer consists of two electrical circuits interlinked by a magnetic circuit. One circuit is designated as the primary circuit and the other as the secondary circuit. The link connecting the two circuits is the core. The primary circuit receives the energy and is referred to as the input. The secondary circuit discharges the energy and is referred to as the output. A transformer is a very efficient device. Excluding the core and copper losses, which are usually about 1%, input power of a transformer equals output power.

Electric Current and Magnetic Field Relationships

In 1819, it was discovered that when an electric current flows in a copper wire (conductor), a magnetic field exists in the space around the conductor. A fixed relationship exists between the direction of the current in a conductor and the magnitude and direction of the resulting magnetic field. See Figure 1-1.

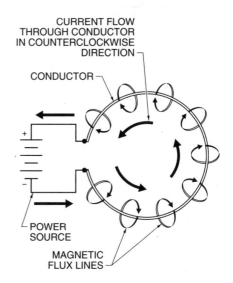

Figure 1-1. A magnetic field exists around a conductor when an electric current flows through the conductor.

A conductor is connected in series with a battery. The battery sends a current through the conductor in a counterclockwise direction. The magnetic flux surround the conductor in the form of complete curves. This is also the case when a straight wire moves some distance from other current-carrying conductors.

Right-Hand Screw Rule. A definite relationship exists between the direction of the current in a conductor and the direction of the magnetic flux surrounding the conductor. The *right-hand screw rule* states that the direction of advance of a right-hand screw corresponds with the direction of current flow through a conductor. The direction in which the screw is rotated corresponds with the direction of the magnetic flux surrounding the conductor. See Figure 1-2. In a conductor, when the right hand is used with the thumb pointing in the direction of the current flow, the fingers point in the direction of the magnetic flux surrounding the conductor. The magnetic flux produced by the current flow through a conductor consists of circular magnetic lines flowing in a clockwise direction.

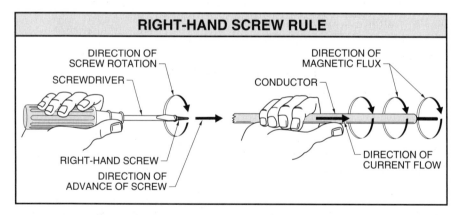

RIGHT-HAND SCREW RULE

DIRECTION OF SCREW ROTATION

SCREWDRIVER

RIGHT-HAND SCREW

DIRECTION OF ADVANCE OF SCREW

DIRECTION OF MAGNETIC FLUX

CONDUCTOR

DIRECTION OF CURRENT FLOW

Figure 1-2. The direction of advance of a right-hand screw corresponds with the direction of current flow through a conductor. The direction in which the screw is rotated corresponds with the direction of the magnetic flux surrounding the conductor.

When two conductors are placed close to each other, the fluxes combine to increase the flux density between the two conductors. See Figure 1-3. The current is flowing in opposite directions in each conductor.

Transformers

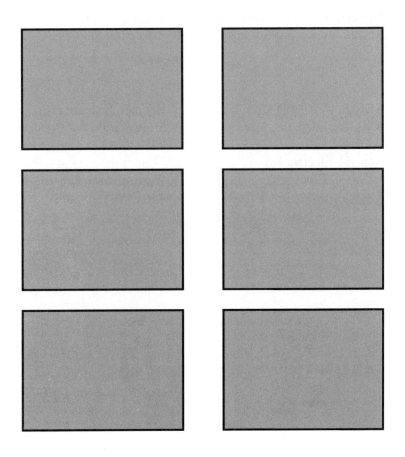

Kenneth R. Edwards

AMERICAN TECHNICAL PUBLISHERS, INC.
HOMEWOOD, ILLINOIS 60430

ACKNOWLEDGMENTS

The author and publisher are grateful to the following companies and organizations for providing technical information and assistance.

ABB Power T&D Company Inc.
Allis Chalmers
Cooper Industries
General Electric Co.
Jefferson Electric Co.

Kemper Insurance Co.
Minnkota Power Cooperative, Inc.
United States Department of the Interior
Westinghouse Electric Corp.

1 2 3 4 5 6 7 8 9 – 96 – 9 8 7 6 5 4 3 2 1

Printed in the United States of America

ISBN 0-8269-1603-1

CONTENTS

INTRODUCTION

Transformers presents basic electrical and magnetic principles as applied to transformers as well as the advanced principles of transformer operation. Vector diagrams are provided which illustrate the current, voltage, and flux relationship in transformers. Transformer classification is explained with respect to service, purpose, installation requirements, and phase. Transformer cooling methods are discussed along with the accessories included on power transformers to enable safe operation of transformers. Additional chapters contain information on tap changers, transformer connections, autotransformers, and reactors.

Transformer maintenance and inspection requirements are presented, including an Inspection Checklist and the latest safety rules that must be followed when working with transformers. The Appendix contains many useful tables and other related information. The expanded Glossary provides quick access to the definitions of technical terms.

The Publisher

1 TRANSFORMER PRINCIPLES

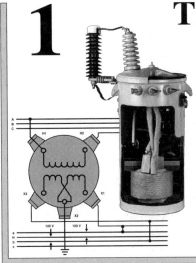

Alternating current can be more economically transmitted over long distances than direct current. AC voltages may be stepped up to a high voltage or stepped down to a low voltage through the use of transformers. High-voltage transmission makes it possible to generate electrical energy more cheaply in large, efficient central generating stations and transmit it considerable distances.

POWER DISTRIBUTION

The success of the transformer is due to George Westinghouse and William Stanley. Westinghouse secured two English patents in 1885 and turned them over to Stanley to develop. In 1886, Stanley succeeded in lighting some stores in Great Barrington, MA. The transformer he designed was capable of stepping down 500 V on a line from a generator a half-mile away.

Increase of Voltage

Transformers are used in the transmission and distribution of electrical power to reduce the size of the conductors carrying the power. Electrical power is the product of voltage and current. To calculate power, apply the formula:

$P = E \times I$

where

P = power (in W)

E = electrical pressure (in V)

I = electric current (in A)

1

Example: Calculating Power

What is the power in a 120 V circuit carrying 1 A?

$$P = E \times I$$
$$P = 120 \times 1$$
$$P = \textbf{120 W}$$

The electrical power remains the same if the circuit is carrying 1 A at 100 V or 10 A at 10 V. For example, 100 V × 1 A = 100 W and 10 V × 10 A = 100 W.

The size of the conductor necessary for transmitting a given amount of power a given distance depends inversely on the square of the voltage. If the voltage between transmission lines is doubled, a conductor only $\frac{1}{4}$ as large is necessary to transmit the same power over the same distance with the same line loss. If the voltage is increased three times, the size of the conductor is only $\frac{1}{9}$ as large.

In the transmission of electrical power, a transformer is used to raise (step up) the voltage at the generator to a high value, reducing the size of the transmission conductor. The installation and maintenance costs are less for a small conductor.

AC generators (alternators) normally produce voltages of up to 22,000 V which are stepped up to a higher voltage for economical transmission to the load or to the place where the power is used. Transmission lines usually operate at voltages from 7200 V to 69,000 V for transmitting industrial power short distances or for low power demand areas, such as residential areas. Very high voltages such as 1,300,000 V are necessary to transmit large amounts of power over considerable distances.

Reduction of Voltage

High voltages are required for economical transmission. Such voltages are undesirable when electrical power is distributed among buildings or within densely populated areas. Transformers lower (step down) the voltage at the load or building where the power is required to a value that is safe and economical to use inside residential, commercial, or industrial buildings. For residential wiring, the normal voltage is 115/230 V. The voltage in a commercial or industrial building is usually 120/208 V for appliances and incandescent lighting and 277/480 V for fluorescent lighting and power.

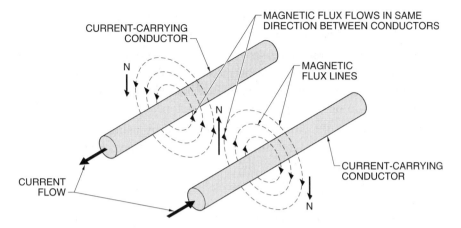

Figure 1-3. Two conductors placed close to each other increase the flux density between the conductors.

When the right-hand screw rule is applied to each conductor, the magnetic flux from each conductor flows in the same direction between the two conductors. Current flowing around a coil in a counterclockwise direction produces a north magnetic pole at the top of the coil.

Magnetic Flux

A magnetomotive force must be applied to establish magnetic flux in a magnetic circuit. *Magnetomotive force (MMF)* is the magnetic pressure that determines the number of flux lines through a medium. This force may come from the magnetic field around a permanent magnet or from the flow of electric current through a conductor or coil of wire. The unit of measure of magnetomotive force is the gilbert. The flux lines always form closed loops and travel in a path known as the magnetic circuit. The magnetic field is represented by the flux lines. The total number of flux lines are the magnetic flux. Magnetic flux is measured in maxwells. One maxwell is equal to one magnetic flux line.

The magnetomotive force produces the magnetic flux. The magnetizing force is the magnetomotive force that produces the magnetic flux per unit length of the magnetic circuit. An *oersted* is the magnetomotive force per unit length. The magnetizing force is found by dividing the magnetomotive force (in gilberts) by the length of the magnetic circuit (in centimeters). The *gilbert* is one oersted-centimeter.

A *maxwell* is the unit of measurement of the total number of flux lines in a magnetic field. *Flux density* is the measure of the flux lines per unit area taken at right angles to the direction of flux. Its unit of measure is the gauss, which is equal to one maxwell per square centimeter.

The direction of the flux lines may be found by using a compass. The needle always points in the direction of the flux lines. See Figure 1-4. Concentric rings are formed by the iron filings around the conductor. The amount of force in each path decreases as the length of the path increases. The flux density is strongest next to the wire and is weakest farthest from the wire.

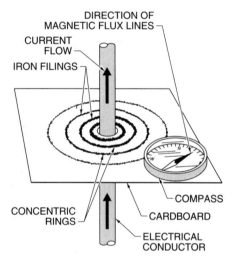

Figure 1-4. The direction of flux lines may be found by using a compass.

Magnetic Conductivity (Permeability)

Different media allow magnetic flux lines to pass through them with different degrees of ease. For example, air, copper, and zinc are not good conductors of magnetic flux lines. Iron, however, is an extremely good conductor of magnetic flux lines and is employed in place of air or other substances whenever possible. Iron, used instead of air as a medium for conducting magnetic flux, increases the number of magnetic flux lines through a coil by over 2000 times.

For example, a single loop of wire with a certain current might produce 200 flux lines with no iron near the arrangement. If a soft-iron core is introduced so that it passes through the circular loop, the number of magnetic lines increases to 400,000 lines ($200 \times 2000 = 400,000$) without any change in the amount of current in the electrical circuit.

Magnetic permeability is the measure of the ease with which certain substances conduct magnetic flux lines. The Greek letter μ is often used to denote permeability. While iron is the best conductor of magnetic flux lines, the various grades of iron differ greatly in magnetic conducting ability. The softer the iron, the greater its conductivity (permeability). Pure soft iron is used in the cores of most transformers.

Retentivity

The magnetic flux lines disappear when the current is removed from a circuit. If the circuit is looped around an iron core when the current is stopped, a considerable number of magnetic lines remain in the iron. *Retentivity* is the ability of a substance to retain magnetic flux lines after the magnetizing force has been removed. The harder the iron, the greater the number of the retained magnetic flux lines. Many lines are retained if hard steel is used as a core instead of soft iron. The hard steel becomes a permanent magnet. Only soft iron should be used in transformer coils because high retentivity values are undesirable in transformers.

Magnetic Induction

Steel is composed of molecules. Each molecule is regarded as a tiny magnet in which its poles are distributed at random in a demagnetized steel bar. Demagnetized steel may be magnetized to a small degree if it is rubbed by a permanent magnet. The same pole of the permanent magnet must be used to rub the demagnetized steel in the same direction each time.

When the demagnetized steel bar is rubbed with the pole of a permanent magnet, some of the molecular magnets respond by lining up end-to-end thus having a combined field that makes the steel bar a weak permanent magnet. *Magnetic induction* is the setting up of flux lines in a material by an electric current. The magnetic lines may be in air, steel, or iron.

For example, when the current in a circular loop is 1 A, a certain number of magnetic lines are produced in the air. Because the unit for electric current (amperes) is not the same as the unit for magnetic lines, the relationship between the number of amperes of current and the number of magnetic lines must be expressed symbolically. To produce a magnetic field of a given number of lines, a current of a certain number of amperes must exist in the circular loop. *Ampere-turns* is a unit of magnetomotive force which is obtained by multiplying the current in amperes by the number of turns in a coil. Thus the magnetic flux lines are directly proportional to the ampere-turns of the coil. The *saturation point* of a magnetic circuit is the point where an increase in magnetizing force (ampere-turns) does not greatly increase the flux density. At the saturation point, it is no longer practical to increase the current or turns of a coil. See Figure 1-5.

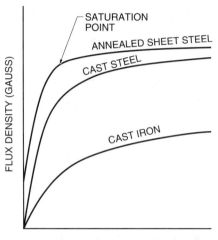

Figure 1-5. At the saturation point, an increase in current or turns of a coil does not significantly increase the magnetizing force.

Electromotive Force of Induction

Magnetic induction is the setting up of magnetic flux lines by an electric current. Given the proper conditions, a changing magnetic field can induce an electric current. See Figure 1-6.

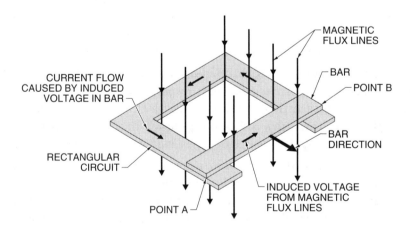

Figure 1-6. Electric current can be induced by magnetic flux lines.

A bar moved back and forth along a rectangular circuit, through the flux lines of a magnetic field picks up an induced voltage from the rectangular circuit. If the bar is moved to the right, an induced voltage flows from point A to point B in the bar. An induced voltage flowing from point A to point B in the bar causes a current to flow through the closed part of the rectangular circuit in a counterclockwise direction. Moving the bar further to the right causes more and more flux lines to be introduced into the closed part of the rectangular circuit.

Right-Hand Rule

The *right-hand rule* is the relationship between the movement of an inductor and the induced voltage in the inductor as it cuts magnetic flux lines. The right-hand rule indicates the direction of induced voltage. With the right-hand rule, the right hand is held so that the thumb, index finger, and middle finger are perpendicular to one another. Placing the hand in a magnetic field so that the index finger points in the direction of the magnetic flux lines and the thumb points in the direction in which an inductor is moved across the lines of force, the middle finger points in the direction of the induced voltage. See Figure 1-7.

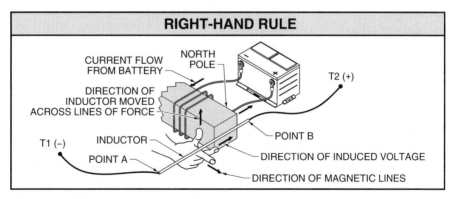

Figure 1-7. The right-hand rule indicates the direction of induced voltage.

The magnetic flux lines flow from the north pole, as shown by the index finger. The lines are produced by a current from the battery passing around the pole in a counterclockwise direction. If the inductor moves in the direction indicated by the thumb, the induced voltage is from point A to point B as indicated by the middle finger.

Relative Motion Between Circuit and Magnetic Field

An electrical voltage can only be induced while there is a relative motion between a conductor or a circuit and a magnetic field. There can be no induced voltage if the magnetic field that flows through a circuit has a steady value and the circuit is stationary. A direct current that does not vary in its amount from one second to the next cannot be employed in transformers to effect changes in voltages.

No current can flow in a circuit if the circuit, through which the number of magnetic flux lines is changing, is an open circuit. Induced voltage from the magnetic flux lines is present in the bar moving across the open circuit. See Figure 1-8.

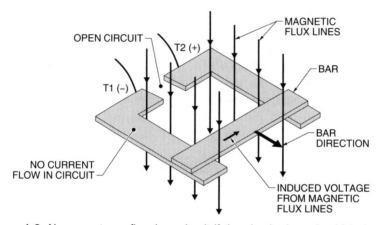

Figure 1-8. No current can flow in a circuit if the circuit, through which the number of magnetic flux lines is changing, is an open circuit.

The voltage exists in the portions of the circuit and according to the right-hand rule, T2 is positive with respect to T1. If, while the bar is moving and the voltage is being induced, T1 and T2 are brought together, thus closing the circuit, the induced current can flow.

Left-Hand Rule

In the design of transformers the left-hand rule is as important as the right-hand rule. The *left-hand rule* is the relationship between the current in a conductor and the magnetic field existing around the conductor. With the left-hand rule, the left hand is held so that the thumb, index finger, and middle finger are perpendicular to one another. Placing the hand in a magnetic field so that the index finger points in the direction of the flux lines and the middle finger points in the direction of the current in a conductor placed in the field perpendicular to the lines of force, the thumb points in the direction in which the conductor tends to move. See Figure 1-9.

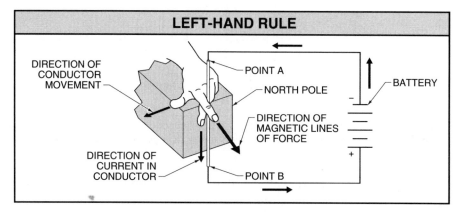

Figure 1-9. The left-hand rule is the relationship between the current in a conductor and the magnetic field existing around the conductor.

This is shown when a bar is allowed to slide along two parallel conductors that have magnetic flux lines flowing in the downward direction. A battery sends current through the circuit in a counterclockwise direction. According to the left-hand rule, the sliding bar tends to move to the left. This motion removes flux lines from the circuit. See Figure 1-10.

Counter Electromotive Force

According to the right-hand rule, the induced voltage is clockwise around a conductor moving across a magnetic field whenever the number of flux lines flowing through a circuit is decreased. If the current in the movable conductor is in such a direction and of such an amount as to cause the conductor to actually move to the left (left-hand rule), thus reducing the number of flux lines, a voltage is simultaneously induced in the circuit in a clockwise direction or directly opposite to that of the applied voltage.

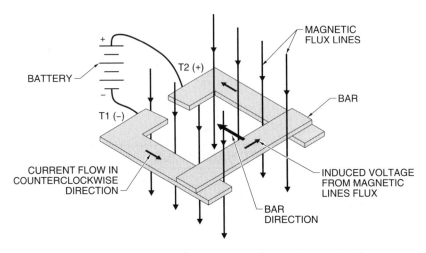

BATTERY

T2 (+)

T1 (-)

MAGNETIC
FLUX LINES

BAR

INDUCED VOLTAGE
FROM MAGNETIC
LINES FLUX

CURRENT FLOW IN
COUNTERCLOCKWISE
DIRECTION

BAR
DIRECTION

Figure 1-10. A bar allowed to slide along two parallel conductors that have magnetic flux lines flowing in the downward direction tends to move to the left according to the left-hand rule.

The clockwise induced voltage, due to the motion of the conductor, sends a current around the circuit in a clockwise direction. The resultant current is the difference between the current produced by the power supply and that resulting from the induced voltage. Whenever the conductor stops moving, there is no induced voltage nor current, and the current in the circuit may be expressed by Ohm's law. To calculate current in a circuit containing neither induced voltage nor current, apply the formula:

$$I = \frac{E}{R}$$

where

I = current in circuit (in A)

E = voltage applied to circuit (in V)

R = resistance of circuit (in Ω)

Example: Calculating Current–No Induced Voltage

What is the current flowing in a 200 V circuit that has 1 Ω of resistance?

$$I = \frac{E}{R}$$

$$I = \frac{200}{1}$$

$$I = \textbf{200 A}$$

When the conductor is moving to the left, the current in the circuit is less than when the conductor is stationary due to the induced counter electromotive force (counter EMF). To calculate current when the conductor is in motion, apply the formula:

$$I' = \frac{E - e}{R}$$

where

I' = current in circuit with counter EMF (in A)

E = applied voltage (in V)

e = induced voltage (in V)

R = total resistance (in Ω)

Example: Calculating Current–Induced Voltage Present

What is the current in a 200 V circuit that has 1 Ω of resistance and 100 V of counter EMF?

$$I' = \frac{E - e}{R}$$

$$I' = \frac{200 - 100}{1}$$

$$I' = \frac{100}{1}$$

$$I' = \mathbf{100\ A}$$

The induced voltage (e) is directly opposite in direction to the applied battery voltage. This voltage is counter EMF because of its opposite direction. The counter EMF is produced by induction or relative motion between an inductor and a magnetic field and is referred to as the EMF of induction.

If the direction of the current around the circuit is reversed, the movable portion of the conductor moves to the right, according to the left-hand rule. In this case, the number of flux lines flowing through the circuit is increased if the movable conductor moves. The resulting induced voltage is counterclockwise, according to the right-hand rule, in a direction exactly opposite to that of the applied voltage from the power supply.

Introducing flux lines into a circuit produces an induced voltage equal in value to, but in opposite direction to the voltage produced by removing the same number of flux lines in the same length of time.

Rate of Change in Magnetic Flux

If an inductor moves across a circuit at a certain rate, the number of flux lines flowing through the circuit is changed in exact proportion to that rate. If the distance moved by an inductor is 10 feet per second (ft/sec) or 120″, its velocity is expressed as feet or inches per second.

If the length of the inductor is 10″ and its velocity is 120 inches per second (in./sec), it sweeps over an area of 1200 sq in. ($120 \times 10 = 1200$). If the number of magnetic flux lines per square inch flowing downward through a circuit is 20,000, the number of lines traversed by the moving inductor during one second of its motion, is 24,000,000 ($1200 \times 20,000 = 24,000,000$).

The total area swept over by the moving portion of any circuit, multiplied by the magnetic density (number of flux lines per unit area), gives the number of lines introduced to or removed from the circuit. The induced voltage in any circuit, due to the rate of change in the magnetic flux lines induced by the circuit is obtained by dividing the number of flux lines cut in one second by 100,000,000.

According to Lenz's law, it is necessary for 100,000,000 (10^8) magnetic flux lines to be cut in one second to produce 1 V. If the number of magnetic flux lines per square inch, traversed by a moving inductor during one second of its motion is 24,000,000, the voltage produced by the conductor equals .24 V ($24,000,000 \div 100,000,000 = .24$).

Rate of Change in Current

If an inductor remains stationary and if 24,000,000 flux lines are uniformly removed each second from flowing through the circuit, the voltage induced in the circuit is exactly the same as though the conductor were moved across the circuit at the rate of 120 in./sec, (if its length between the portions of the circuit upon which it slides is 10″ and if the magnetic density is 20,000 flux lines per square inch). Induced voltage always depends on the relative motion of any circuit and a magnetic field flowing through the circuit. The number of lines flowing through the circuit may be removed by reducing the amount of current used to produce the field.

Assume that a variable resistor (rheostat) is connected into the circuit. See Figure 1-11. By moving the variable contact to the right, the resistance at R_x increases, the current in the circuit decreases, and the number of flux lines flowing through the circuit is reduced. The quicker the contact is moved, the greater the rate of change in the current and the greater the rate of change in the number of flux lines. If the circuit is suddenly opened, the flow of the current is stopped and the number of flux lines is reduced to zero, provided there is no iron in or near the circuit.

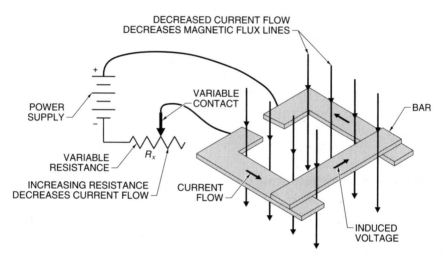

Figure 1-11. The number of lines flowing through the circuit may be removed by reducing the amount of current used to produce the field.

Transformer Operation

A simple transformer may be developed from a round bundle of soft iron conductors, forming the core of a coil of wire consisting of three turns which has a current flow through it in a clockwise direction from a battery. See Figure 1-12. This causes the upper end of the core to be the south pole and the magnetic flux lines to have a downward flow. The circuit ABCD is arranged so that a certain number of magnetic flux lines flow downward through it. No current flows in circuit ABCD as long as the parts of the arrangement are stationary and the battery current does not change.

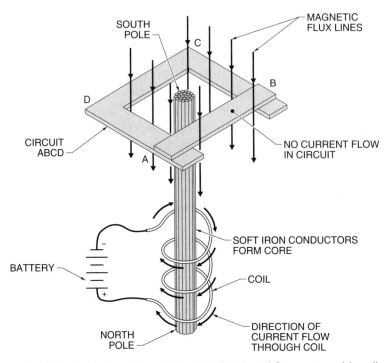

SOUTH POLE

MAGNETIC FLUX LINES

C

B

D

CIRCUIT ABCD

NO CURRENT FLOW IN CIRCUIT

A

SOFT IRON CONDUCTORS FORM CORE

BATTERY

COIL

NORTH POLE

DIRECTION OF CURRENT FLOW THROUGH COIL

Figure 1-12. A simple transformer may be developed from a round bundle of soft iron conductors and a coil of wire.

If the coil circuit is opened and the magnetizing current in the coil surrounding the iron core drops to zero, the number of flux lines flowing downward through circuit ABCD decreases rapidly and a current is induced in circuit ABCD in a clockwise direction. See Figure 1-13. The induced current in circuit ABCD is in the same direction as the current in the coil surrounding the core that produced the magnetic flux lines that flow through circuit ABCD. The coil surrounding the core may be designated as the primary, and circuit ABCD as the secondary of a transformer. The two electrical circuits are connected (interlinked) by the magnetic circuit consisting of the flux lines flowing through the iron core, the primary, and the secondary. If the battery circuit is closed by joining T1 and T2, the current through the coil increases, increasing the number of flux lines flowing through the secondary (circuit ABCD), and a counterclockwise current is produced in the secondary.

Magnetic Leakage

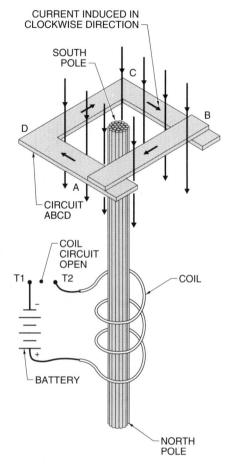

Not all the magnetic flux lines produced by the magnetizing force of the ampere turns of the primary pass through the secondary. The shape of the magnetic field produced by the magnetizing force of the primary coil tends to arc outward from the coil. See Figure 1-14. *Leakage lines* are the magnetic lines produced by the primary coil which do not pass through the secondary. The greater the number of leakage lines, the less useful effect realized.

The primary and secondary must be arranged so that as many of the magnetic flux lines as possible flow through both the primary and secondary. This is accomplished by making the magnetic circuit (core) a complete, circular, closed circuit of iron, and winding the primary on this in close turns. See Figure 1-15.

Figure 1-13. Current is induced in circuit ABCD in a clockwise direction if the coil circuit is opened and the magnetizing current in the coil surrounding the iron core drops to zero.

The secondary may be wound over the primary. This is the most magnetically efficient form of a transformer. There are no joints to interrupt the passage of the magnetic flux lines, and the path of the magnetic circuit is circular. The disadvantage of this form is the increased labor required in its construction. Each turn of the primary and secondary must be threaded through the center hole by hand. Labor is considerable if there are many turns to be wound on the core.

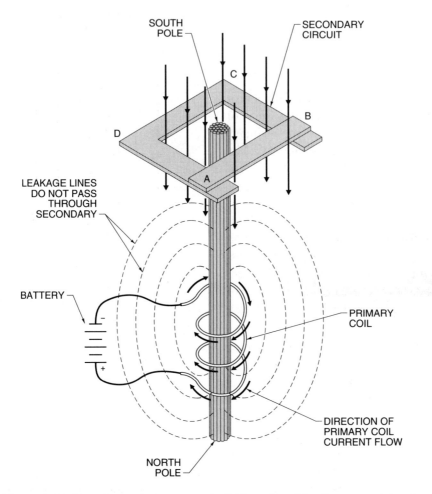

Figure 1-14. The shape of the magnetic field produced by the magnetizing force of the primary coil tends to arc outward from the coil.

Open- and Closed-Core Transformers

An *open-core transformer* is a transformer having a magnetic circuit consisting partly of air and partly of iron. Open-core transformers are desirable in the operation of certain telecommunication and electronic equipment. A *closed-core transformer* is a transformer having a magnetic circuit consisting completely of iron. Most classes of power service employ closed-core transformers.

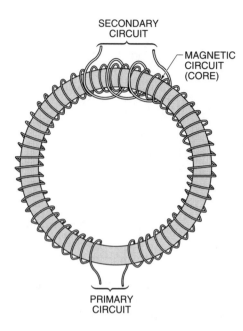

ABB Power T&D Company Inc.

Figure 1-15. The primary and secondary of a transformer are arranged so that the magnetic circuit (core) is a complete, circular, closed circuit of iron, and the primary is wound around it in close turns.

Effect of Turns in Primary

Increasing the number of turns in the primary windings increases the possible magnetizing force due to the primary ampere turns. This increases the amount of the magnetic flux with a given primary current.

Effects of Turns in Secondary

If a second circuit (turn) is connected in series with the first turn, the total induced voltage of the secondary is twice the induced voltage of the circuit with a single turn. See Figure 1-16. The secondary voltage depends on the number of turns wound to constitute the secondary coils. This holds true for any arrangement of turns.

Transformer Ratio (Turns Ratio)

Turns ratio is the ratio between the voltage and the number of turns on the primary and secondary windings of a transformer. The relationship of the voltage produced in the secondary winding of a transformer to the primary depends on whether the secondary winding has the same number of turns, a fewer number of turns, or a greater number of turns. To calculate turns ratio, apply the formula:

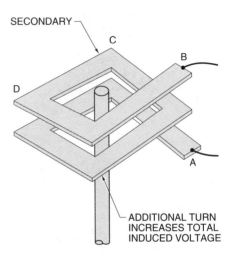

Figure 1-16. The total induced voltage of a secondary with two turns is twice the induced voltage of the circuit with a single turn.

$$\frac{N_p}{N_s} = \frac{E_p}{E_s}$$

where

N_p = number of turns in primary

N_s = number of turns in secondary

E_p = primary voltage (in V)

E_s = secondary voltage (in V)

The secondary voltage is higher if the secondary has a greater number of turns than the primary. The secondary voltage is lower if the secondary has a fewer number of turns than the primary. If the secondary has one-half as many turns as the primary, the secondary voltage is approximately one-half that of the primary. A *step-down transformer* is a transformer in which the primary voltage winding has a greater number of turns than the secondary winding. A *step-up transformer* is a transformer in which the secondary voltage winding has a greater number of turns than the primary winding. In a step-up transformer, the voltage is stepped up.

Unless otherwise specified, the transformer ratio is the turns ratio. In practice, it is customary to specify the ratio of transformation by writing the primary number first. The primary winding is the winding which receives the energy, and it is not always the high-voltage winding. Strict adherence to this policy eliminates the necessity of adding the terms step-up and step-down to a transformer. For example, a transformer with a ratio of 10:1 is a step-down transformer, and a transformer with a ratio of 1:10 is a step-up transformer. Although step-up and step-down refer to the voltage value, the transformer ratio may also be referred to as the turns ratio because turns ratios vary in the same proportion.

An inverse ratio exists between the number of turns, the voltage, and the current in the primary and secondary of a transformer. To determine the inverse ratio, apply the formula:

$$\frac{N_p}{N_s} = \frac{I_s}{I_p} \quad or \quad \frac{E_p}{E_s} = \frac{I_s}{I_p}$$

where

N_p = number of turns in primary

N_s = number of turns in secondary

E_p = primary voltage (in V)

E_s = secondary voltage (in V)

I_p = primary current (in A)

I_s = secondary current (in A)

Transformer Losses

The energy losses in a transformer may be classified as the primary loss, the secondary loss, and the core loss. The primary loss is caused by the resistance in the primary winding and is expressed in watts. The secondary loss is caused by the resistance in the secondary windings and is expressed in watts. When there is no load on the transformer, the transformer is not supplying current, and the power loss in the secondary is zero due to a very small current. The loss in the iron core is a power loss and is expressed in watts. The core loss may be divided into eddy-current loss and hysteresis loss.

Eddy-Current Loss. Eddy current in an iron core is a true electric current induced in the iron by the changing magnetic flux, just as if the iron were the desired conductor. Induced currents in an iron core are undesirable because such currents require power input to the device and convert the electrical power into heat that is not only unavailable for useful electrical output, but tends to heat the primary and the secondary windings, thus increasing the resistance and the power loss in both coils.

Eddy-current loss may be reduced if the iron core is constructed of thin sheets (laminations) of soft iron which are insulated from each other. See Figure 1-17. If the core is made of a solid piece of iron, the eddy current circulates throughout the entire core. If the core is divided into two parts, there is only half the current in each part, and the loss in each core is only $\frac{1}{4}$ as much as the loss in the solid iron core. A laminated core divided into seven parts has only $\frac{1}{7}$ of the current present in each part. The loss in the core is $\frac{1}{49}$ as much as the loss in a solid iron core.

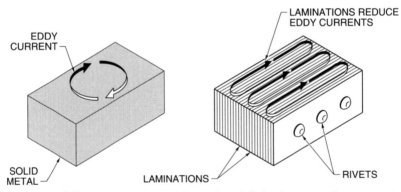

Figure 1-17. Eddy-current loss may be reduced if the iron core is constructed of thin sheets of soft iron which are insulated from each other.

Hysteresis Loss. Hysteresis is a Greek word meaning to lag. *Hysteresis* in an iron core is the magnetic flux that lags behind the magnetizing force that causes them. The nature of the hysteresis effect may be understood by considering that whenever a new piece of iron is subjected to a magnetizing force and the magnetizing force is removed, a portion of the magnetic flux remains in the iron. To remove this residual magnetism, another magnetizing force must be applied to the iron in a direction opposite to that of the initial magnetizing force. Energy has to be supplied to demagnetize the iron.

In the operation of a transformer, the primary current is constantly changing in value and direction so that the magnetizing force of the primary is first in one direction and then in the opposite direction. For each change in magnetization from one direction to the opposite, a certain amount of power or energy is demanded. The total amount of energy required per second depends on the number of reversals that occur each second and the quality of the iron. Soft iron retains few flux lines and requires less applied energy to be demagnetized.

In an alternating waveform, the value increases from zero to a certain positive maximum, decreases to zero, increases in the opposite direction to a negative maximum, and returns to zero. The value may be alternating voltage, current, or magnetic flux. One complete cycle of changes in value is one complete waveform. See Figure 1-18. The frequency of the waveform is 60 Hz if there are 60 complete cycles per second.

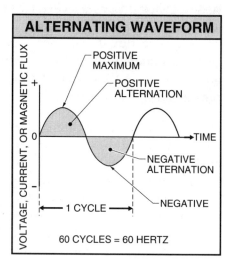

Figure 1-18. In an alternating waveform, the value increases from zero to a certain positive maximum, decreases to zero, increases in the opposite direction to a negative maximum, and returns to zero.

Because magnetization affects the molecular constitution of the iron of the core, changing the position of the molecules from pointing in a certain direction to pointing in the opposite direction has practically the same effect as the rapid hammering of the iron, which produces heat. The core loss of a transformer varies as the square of the load current. The core loss depends on the terminal voltage and the frequency of the supply. The core loss is constant from no load to full load because the factors affecting it (frequency, voltage, etc.) are also constant.

Transformer Types

Transformers are classified into three types according to their relative disposition of iron and copper; the core, H (distributed), and shell types.

Each transformer type possesses characteristics which specially adapt it to certain conditions. The core type has a relatively light core of small sectional area and a long magnetic circuit. The copper is relatively heavy containing more turns but has a shorter length. The core type is more easily wound as cylindrical formed coils, and the coils are more accessible and expose more surface to radiation. The core type with its relatively large winding space, is better adapted for high voltages which require many turns and heavy insulation. The core type requires small wire with low magnetic flux density because it handles small currents. The shell type is particularly suited for transformers of moderate voltage, requiring few turns and little insulation, large currents, and low frequency with corresponding magnetic flux.

Core. A complete single-phase (1φ) transformer core is built up of sheet steel strips. See Figure 1-19. The legs of the core are the two upright portions on which the coils are placed. The short horizontal parts of the core that do not have windings on them are the yokes. The yokes complete the magnetic circuit and are just long enough to provide room for the coils. The yokes are formed from loose sheet steel strips.

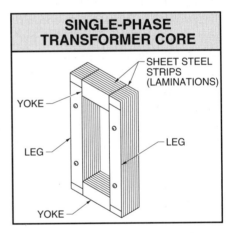

Figure 1-19. A complete 1φ transformer core is built up of sheet steel strips.

The upper yoke is removable to permit the coils to be placed into position on the core. The legs of the core are wrapped with a thick layer of insulating material to prevent electrical contact between the wire of the coil and the iron of the core. See Figure 1-20.

The core of the transformer is constructed of cold-rolled, grain-oriented, sheet-steel laminations. Each lamination is coated on each side with an insulating material to reduce the eddy-current losses.

The two high-voltage coil leads are connected in series and connected to the high-voltage bushing terminals. The two low-voltage windings are connected in series by connecting the second and third secondary coil leads to the center low-voltage terminal. The two outside low-voltage leads are connected to the two outside low-voltage terminals. This connection is used to form an Edison three-wire 1φ system having voltages between the middle terminal and either one of the outside terminals of 120 V and 240 V, between the two outside low-voltage terminals.

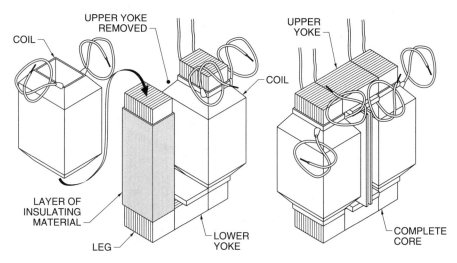

Figure 1-20. The legs of the core are wrapped with a thick layer of insulating material to prevent electrical contact between the wire of the coil and the iron of the core.

Normally on small, 1φ transformers, the low-voltage winding is next to the core, and the high-voltage winding is placed over the low-voltage winding with a layer of insulating material between them. By placing the windings in such a manner, the insulation required for the high-voltage winding is kept to a minimum.

The sheet steel laminations and the coil are held firmly in place by angle or channel braces and bolts (core clamp brackets). The complete transformer is mounted in a welded, pressed-steel tank. The mechanical construction of a transformer varies, depending on the voltage and size. In high-voltage transformers, it is customary to use a tall case and immerse the terminal blocks in oil. See Figure 1-21.

H (Distributed). The H (distributed) type is a modified form of the standard type of transformer. See Figure 1-22. A large rectangular core is divided into four smaller cores with the magnetic circuit of each core in parallel. The core is cross-shaped when viewed from the top. In this type of core, the leakage flux is very small as the coil windings are wound on a center leg and surrounded by the four outside legs of the core structure. There are two low-voltage windings. The high-voltage windings are placed between the two low-voltage windings. This coil arrangement keeps the insulation requirements to a minimum, with high-voltage insulation required only between the high-voltage and low-voltage windings. This method has been used extensively in the construction of distribution transformers.

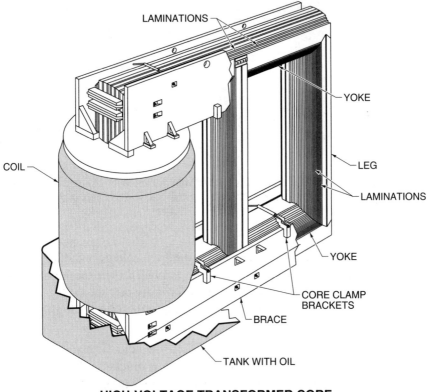

HIGH-VOLTAGE TRANSFORMER CORE

Figure 1-21. The sheet steel laminations and the coil are held firmly in place by angle or channel braces and bolts (core clamp brackets).

The space between the primary and secondary coils allows a circulation of oil upward from the bottom of the case. This aids in conducting the heat produced by the energy losses in the transformer winding to the oil and to the outer case where it is radiated to the outside air.

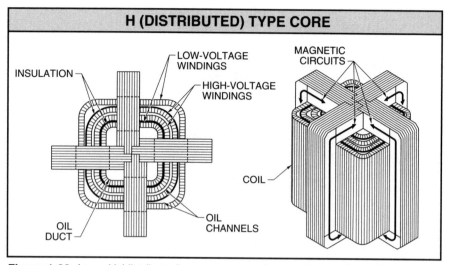

Figure 1-22. In an H (distributed) type core transformer the core is divided into four smaller cores with the magnetic circuit of each core in parallel.

Oil is a better heat-conducting medium than air. Oil carries heat from a transformer to the containing case much better than air. A transformer in oil operates at a much lower temperature. Oil preserves the insulation, keeping it soft and pliable. Oil also prevents oxidation in the air. Consequently, its use is advantageous in producing proper conditions to maintain a uniform core loss and superior insulation. Oil is not permanently damaged by a puncture caused by lightning because the resistance of the oil is only momentarily broken down as the oil immediately flows into the break and seals the insulation. Askarel and Transil are the names of two special transformer oils that have a high-insulating value and dielectric strength.

Shell. In a shell-type transformer, the core leakage flux is cut to a minimum because the coil windings surround the inner core, and the outer core surrounds the coil. The core is constructed of thin sheet steel strips. See Figure 1-23.

In large shell-type power transformers, the primary and secondary coils are usually wound in pancake form on formers. See Figure 1-24. A flat rectangular copper strip is used with one turn per layer in many layers. Each flat rectangular copper strip is specially insulated. The thin pancake coils are treated with an insulating compound and wound with a number of layers of tape according to the voltage for which they are designed.

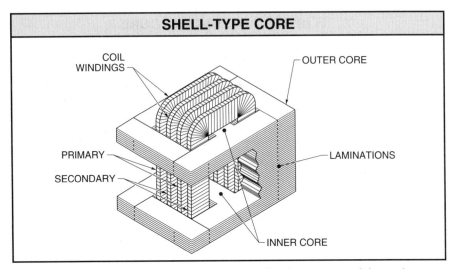

Figure 1-23. In a shell-type core, the leakage flux is cut to a minimum because the coil windings surround the inner core, and the outer core surrounds the coil.

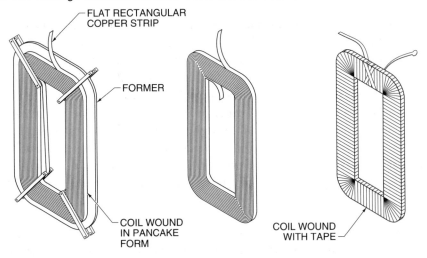

Figure 1-24. Large shell-type power transformers have the primary and secondary coils wound in pancake form on formers.

Each layer is given several coats of insulating varnish baked on in ovens. The coils are then assembled into groups of two or more sections. The groups are then assembled into complete windings, the primary and the secondary being intermixed or sandwiched to reduce magnetic leakage. Suitable insulating barriers are interposed between the various groups. See Figure 1-25.

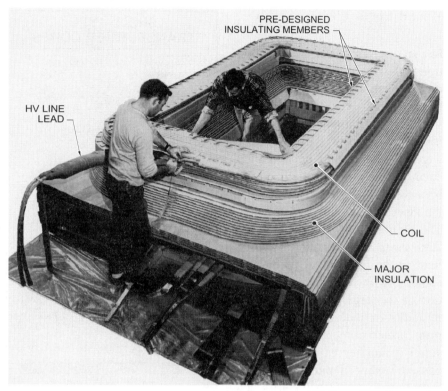

ABB Power T&D Company Inc.

Figure 1-25. Pancake coils are assembled into groups of two or more sections and the groups are assembled into complete windings.

In high-voltage transformers, the insulation is heavily reinforced for a considerable length of the conductor nearest to the terminal leads. The extra dielectric strength of the insulation of these end turns is a safeguard against breakdowns which might occur due to the excessive voltages from lightning discharges or other surges.

Shell-type transformer cores include 3-leg, type D, and 7-leg cores. See Figure 1-26. Single-phase transformers use a 3-leg core design. Small 3φ transformers are generally designed with a type D core. The 7-leg core has the three phases mounted in parallel on a long and narrow configuration. The two auxiliary legs of the 7-leg design provide for symmetry among the three magnetic circuits of the core. This design has a flux sharing characteristic which allows more uniform distribution of flux between the various legs of the core. The more uniform distribution of flux provides a lower content of harmonics.

Wound Cores

In the past, the sheet steel laminations used in core and shell-type transformers were usually cut from large sheets (about 10′ long and 3′ wide) which the steel mills could manufacture easily. These sheets were rolled while red hot to the desired thickness. The rolling process improved the magnetic qualities of the sheet in the direction of its length. The laminations were cut from the large sheets so that the magnetic flux lines in the transformer core would coincide with the lengthwise direction of the large sheet. The number of joints or pieces of steel in the transformer core were kept to as small a number as possible. This caused L-shaped punched laminations to be used extensively for constructing transformer cores.

It was not as easy to cut or punch the L-shaped pieces from a sheet of steel as it was to cut or punch the simple straight strips. It required a careful layout of the L-shaped strips on the large sheet to keep the waste or scrap material to a minimum. The increase in efficiency of a transformer due to the improved magnetic qualities of the iron in the core was obtained only by an increase in cost of construction. This was not an efficient improvement.

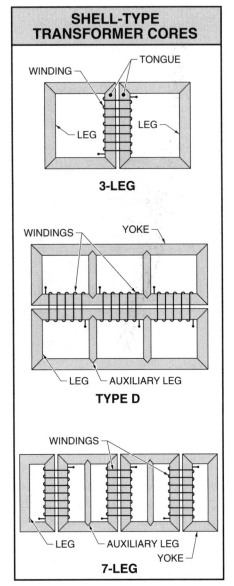

Figure 1-26. Shell-type transformer cores include 3-leg, type D, and 7-leg cores.

Steel manufacturers developed a method of cold rolling long strips of thin steel that have better magnetic qualities lengthwise. These long strips of steel can be cut easily to any desired width. It is more convenient to handle the roll of strip steel, than to handle the large flat sheets.

The spiral core was first used for current transformers where the primary consisted of one single insulated conductor that passed through the center of the core. See Figure 1-27. The secondary was wound over the spiral core, usually by hand. The path of the magnetic flux lines moves in a circular direction through the core. The most difficult part with the spiral core is the placing of the core in the center of the coil. The annealed, laminated core is squared and clamped in shape before it is annealed. The annealing process improves the magnetic qualities of the core and holds the turns of strip sheet steel closer together.

Some manufacturers developed a wound core which consists of a long strip of silicon steel wound in a spiral around the windings. The wound core has several advantages in that the magnetic circuit is relatively short and has a large cross section, the flux path is always along the grain of the steel, and flux leakage is at a minimum. See Figure 1-28.

The wound core is spirally wound from a continuous strip of cold-rolled steel which is cut at every other turn. These cuts permit assembling the core around a prewound coil which passes through both openings. The cuts are made

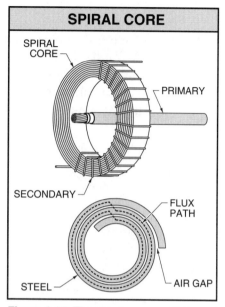

Figure 1-27. The spiral core was first used for current transformers where the primary consisted of a single insulated conductor that passed through the center of the core.

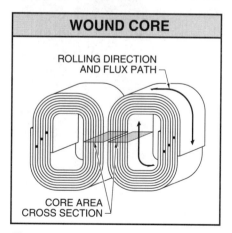

Figure 1-28. A wound core has a relatively short magnetic circuit with a large cross section.

at random positions so that they do not fall in line, which permits the flux to flow between the layers of steel and not across the butt gap. Each sheet must be coated on each side with an insulating material to limit the eddy-current losses.

The Spirakore is a shell type transformer core formerly used by the General Electric Co. for the 25 kVA, 1ɸ, pole-mounted distribution transformer. It is constructed of strips of cold-rolled, grain-oriented steel. A core clamp holds the core in a permanent position. See Figure 1-29. The transformers are installed in pressure-tested, all-welded, oil-filled tanks.

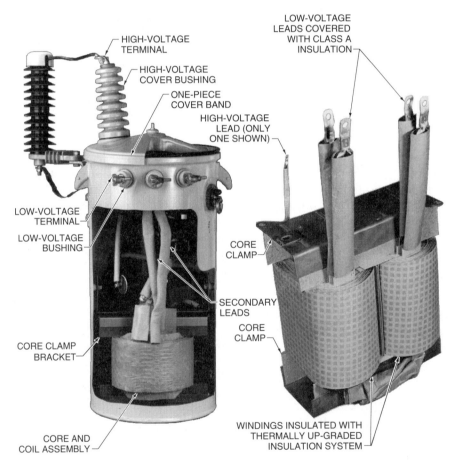

ABB Power T&D Company Inc.   *General Electric Co.*

Figure 1-29. A typical shell core is constructed of cold-rolled, grain-oriented steel.

The wound core used by Asea Brown Boveri uses step-lapped core joints which result in a lower exciting current, a reduction of joint losses, and a lighter transformer with reduced noise. This type of transformer core is referred to as the five-legged core construction. These transformers may be installed in pad-mounted tanks. See Figure 1-30.

WOUND FIVE-LEG
CONSTRUCTION

PAD-MOUNTED
TRANSFORMER

General Electric Co.   *ABB Power T&D Company Inc.*

Figure 1-30. The wound core used by Asea Brown Boveri uses step-lapped core joints which result in a lower exciting current, a reduction of joint losses, and a lighter transformer with reduced noise.

The core of a wound core transformer may be wound on a form using several widths of sheet steel strips. See Figure 1-31. The widest strip is used in the center of the core. This produces a core with a modified or stepped cross section instead of a square cross section.

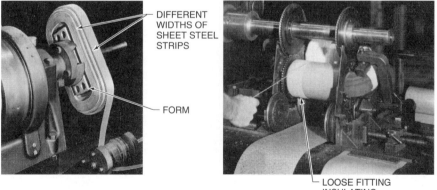

DIFFERENT
WIDTHS OF
SHEET STEEL
STRIPS

FORM

LOOSE FITTING
INSULATING
MATERIAL

Cooper Industries

Figure 1-31. The core of a wound core transformer may be wound on a form using several widths of sheet steel strips.

A loose-fitting cylinder of insulating material is wound around and over the straight sides of the core. The cylinder of insulating material is rotated or turned on the core by a specially built machine. In this way, the primary and secondary coils are wound on the core.

In large distribution and power transformers, four cores may be used with the cores arranged in the center of the coil to give a cross-shaped section. See Figure 1-32. This enables a tall coil of small diameter to be used, which reduces the diameter of the transformer core and also reduces the size of the transformer case. The cores are wound with sheet steel strips of two different widths.

CORE   COIL

Figure 1-32. In large distribution and power transformers, four cores may be used with the cores arranged in the center of the coil to give a cross-shaped section.

Another method of taking advantage of the magnetic properties of a piece of strip sheet steel is to cut the ends of the laminations at an angle of 45° instead of with square ends. See Figure 1-33. The dashed lines show the first layer and the solid lines show the second layer. These laminations are assembled in this order in stacking or building up the core. A wider strip of steel is used for the yoke than is used for the legs, so that angle iron bars can clamp and hold the laminations of the yoke together and in place.

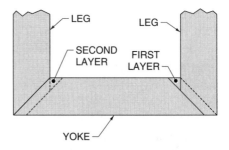

LEG   LEG

SECOND LAYER   FIRST LAYER

YOKE

Figure 1-33. Laminations may be cut at a 45° angle instead of with square ends to take advantage of the magnetic properties of a piece of sheet steel strips.

2 VECTOR DIAGRAMS

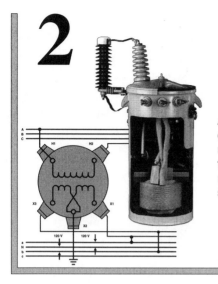

A transformer works on the principle that energy can be efficiently transferred by magnetic induction from one set of coils to another. The current, voltage, and flux relationships in a transformer are represented in a vector diagram.

TRANSFORMER OPERATING PRINCIPLES

Transformers enable the generation of electrical power at a medium-high voltage, the transmittal of the power at a high voltage, and the stepping down of the power to a value suitable for distribution and use by consumers. See Figure 2-1.

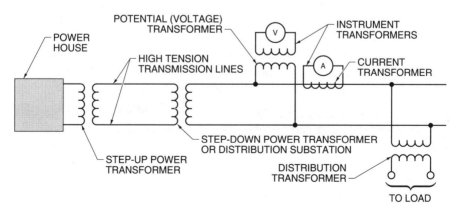

Figure 2-1. Transformers enable efficient transmittal of electrical power as well as other functions.

A transformer works on the principle that energy can be efficiently transferred by magnetic induction from one set of coils to another. This is accomplished by varying the magnetic flux produced by an alternating current (AC). Both sets of coils must be linked together with a magnetic circuit that is common to both coils.

The *primary winding* is the transformer coil which is energized from a source of alternating voltage and current. The *secondary winding* is the transformer coil which delivers the AC and voltage to the load. See Figure 2-2. In some special applications, a transformer may have a third set of coils comprised of a tertiary winding.

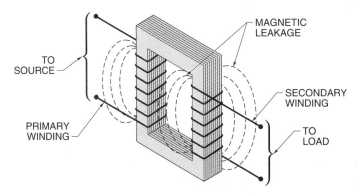

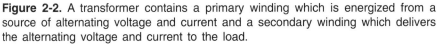

Figure 2-2. A transformer contains a primary winding which is energized from a source of alternating voltage and current and a secondary winding which delivers the alternating voltage and current to the load.

For illustrative purposes, the primary and secondary windings are located on separate legs of the magnetic circuit. This clarifies the understanding of the operation of a transformer. In actual practice, half of the primary and secondary coils are wound on each leg, with sufficient insulation between the two windings and between the windings and the core to insulate the windings from each other and from the core.

A transformer wound with the primary and secondary coils on separate legs functions at reduced efficiency because of increased magnetic leakage. *Magnetic leakage* is the portion of magnetic flux that passes through either one of the coils but not through both. The greater the distance between the two coils, the longer the magnetic circuit and the greater the magnetic leakage. In actual practice, insulated primary and secondary coils are wound on each other or interleaved to keep the magnetic leakage to a minimum.

The primary and secondary windings are often referred to as high-voltage and low-voltage windings. High voltage and low voltage refer to the relative values of the normal voltages of the coils. The operation of a transformer is reversible in the sense that either the high-voltage winding or the low-voltage winding can be used as the primary or secondary. For example, a 115/69 kV transformer may be used to step down 115 kV to 69 kV or may be used to step up 69 kV to 115 kV.

Transformer Operation Without Load

An alternating current flows in a transformer when an alternating voltage is impressed on the primary winding. The AC magnetizes the magnetic (iron) circuit in the positive direction and then in the negative direction. The reversal in the direction of the magnetism is the same as the alternations of the AC supply.

The alternating magnetic flux flowing around the magnetic circuit induces voltages in the primary and secondary coils. The induced voltage per turn of the primary and secondary coils is the same in value and direction at any given instant because both coils are linked by the same magnetic flux.

The induced voltage opposes the voltage impressed across the primary coil. The induced voltage (counter electromotive force) is nearly equal to the impressed voltage. The difference in the two is the voltage necessary to send the current through the primary winding. The current that flows through the primary winding (with no current flowing in the secondary winding) is just enough to establish a magnetic field of sufficient strength to counteract the applied voltage.

The current flowing under these conditions is the exciting current. Because current flows in the primary even when the secondary winding is not connected to a load, it is frequently referred to as the no-load current of the transformer. In the design of a transformer, exciting current must not be permitted to become too large because it is a reactive (wattless) current. *Reactive current* is current out-of-phase with the voltage by 90° and does not represent real energy.

For a given magnetic circuit, the exciting current is decreased by increasing the number of turns in the primary (exciting) winding. The magnitude of the magnetizing force of a coil in which current is flowing is determined by the number of ampere-turns produced by the coil. This is the fundamental principle by which the designer selects the proper ratio of turns and cross-sectional area of the magnetic circuit.

The induced voltage equals the impressed voltage if the primary winding has no resistance, no energy is required in the reversal of the magnetic flux, and if the magnetic circuit has an infinite resistance with respect to the flow of eddy currents so that there is no eddy-current flow. Each turn of the primary winding would have an induced voltage equal to the impressed voltage divided by the number of turns in the primary winding.

Assuming that the secondary coil is so closely interleaved with the primary coil that there is no magnetic leakage lines of flux, in each turn of the secondary coil, the flux sets up a counter electromotive force of the same value as that of each turn in the primary coil. The total induced voltage of the secondary coil is equal to the voltage induced per turn of the primary coil multiplied by the number of turns in the secondary coil. The voltages of the primary and secondary windings are directly proportional to the number of turns in the primary and secondary coils. To calculate turns ratio, apply the formula:

$$\frac{E_p}{E_s} = \frac{N_p}{N_s}$$

where

E_p = induced voltage of primary coil (in V)

E_s = induced voltage of secondary coil (in V)

N_p = total number of turns in primary coil

N_s = total number of turns in secondary coil

This formula may be rearranged to calculate any one value when the other three are known. See Appendix.

Example: Calculating Secondary Voltage

What is the induced secondary voltage of a transformer having a primary winding of 3000 turns and a secondary winding of 300 turns that is connected to a 2400 VAC source?

$$E_s = E_p \times \frac{N_s}{N_p}$$

$$E_s = 2400 \times \frac{300}{3000}$$

$$E_s = 2400 \times .1$$
$$E_s = \textbf{240 V}$$

The voltage induced in the secondary coil equals 240 V and the voltage induced per turn is .8 V (240 V ÷ 300 V = .8 V). The voltage induced in the primary coil equals 2400 V and the voltage induced per turn equals .8 V (2400 V ÷ 3000 V = .8 V).

Because the voltage induced per turn is the same for the primary and secondary coils, the secondary voltage is equal to the primary voltage per turn multiplied by the number of turns in the secondary coil. The secondary voltage equals 240 V (.8 V × 300 V = 240 V).

Transformer Operation With Load

A transformer is loaded when an alternating current is caused to flow through the secondary winding. For example, if a 110 V motor is connected across the 110 V terminals of a 2200 V to 110 V transformer excited from a 2200 VAC source, the motor draws current from the secondary winding of the transformer. The amount of current increases as the power required by the motor to perform its work increases. The 110 V induced voltage of the secondary coil causes the current to flow through the circuit of the motor.

Within the transformer, the secondary current magnetizes the iron circuit in a direction opposite to that of the magnetizing action of the primary exciting current. This magnetizing action lowers the induced electromotive force in the primary and secondary.

The lowered induced electromotive force in the primary and secondary increases the difference between the impressed voltage and the counter electromotive force. This permits a greater current flow in the primary winding. The increase in current flow in the primary winding is the amount that establishes a magnetizing force equal to the magnetizing action of the current flow in the secondary, due to the load. The resultant flux in the core is maintained at a constant value by the primary current, regardless of the value of the load placed on the secondary of the transformer.

Disregarding the small amount of exciting current in the primary coil required to maintain the constant flux in the magnetic circuit, the magnetizing force of the primary winding, due to the load placed on the secondary, is equal to the magnetizing force of the load current. The ampere-turns of the primary is equal to the ampere-turns of the secondary. To calculate ampere-turns, apply the formulas:

$$\frac{I_p}{I_s} = \frac{N_s}{N_p} \quad \text{or} \quad I_p N_p = I_s N_s$$

where

I_p = increase in primary current over no-load exciting current, due to secondary load (in A)

I_s = secondary load current (in A)

N_p = total number of turns in primary coil

N_s = total number of turns in secondary coil

These equations represent the relationship between the currents in the primary and secondary windings because the exciting (no-load) current of a transformer is very small in comparison to its output capacity and may be disregarded for practical purposes. The ratio of the number of turns in the primary to the number of turns in the secondary is equal to the ratio of the primary to the secondary voltages. The ratio of the current in the primary coil to the current in the secondary coil is equal to the ratio of the number of turns in the secondary to the number of turns in the primary. Substituting the ratio E_s/E_p for N_s/N_p *gives the ratio* I_p/I_s = E_s/E_p or $I_p E_p = I_s E_s$.

Neglecting no-load losses, $I_p E_p$ is the volt-ampere energy input of the primary coil. $I_s E_s$ is the energy output of the secondary coil due to the connected load. Thus, the energy is transferred from a primary source of supply having a voltage of E_p volts to the secondary (load side) at a voltage of E_s by the magnetic action of the alternating current in the windings.

Example: Calculating Primary Current

What is the primary current of a 3 kVA rated transformer that has 1760 turns of wire in its primary winding and 88 turns of wire in its secondary winding if ten .5 A lamps are connected to the secondary terminals of the transformer? The primary winding is connected to a 2400 V supply.

1. Calculate secondary voltage.

$$E_s = E_p \times \frac{N_s}{N_p}$$

$$E_s = 2400 \times \frac{88}{1760}$$

$$E_s = 2400 \times .05$$

$$E_s = 120 \text{ V}$$

2. Calculate total secondary current. Total secondary current equals 5 A (10 lamps × .5 A each = 5 A).

3. Calculate primary current.

$$I_p = \frac{N_s}{N_p} \times I_s$$

$$I_p = \frac{88}{1760} \times 5$$

$$I_p = .05 \times 5$$

$$I_p = \mathbf{.25 \text{ A}}$$

The energy delivered to and supplied by the primary winding equals 600 W (2400 V × .25 A = 600 W). The energy delivered to the lamps by the secondary coil is 600 W (120 V × 5 A = 600 W). Each lamp takes .5 A at 120 V. Therefore, each lamp is rated at 60 W (120 V × .5 A = 60 W). Ten lamps require 600 W (60 W × 10 = 600 W).

Because the transformer is rated at 3 kVA (which equals 3000 W with a noninductive load obtained by a pure resistance load such as lamps) and is delivering a load of 600 W, it is loaded at 20% (600 W ÷ 3000 W × 100 = 20%) of its rated capacity. Five times as many lamps of the same rating may be connected without exceeding the rating of the transformer. In such an application, the current in the secondary is 25 A (50 lamps × .5 A = 25 A), and the power delivered by the secondary is 3000 W (25 A × 120 V = 3000 W). The current in the primary is 1.25 A (.25 A × 5 = 1.25 A) and the power delivered to the primary is 3000 W (2400 V × 1.25 A = 3000 W).

This transfer of energy from primary to secondary without any loss represents the ideal condition of an ideal transformer. In practical applications, such a condition cannot exist because:

- The primary and secondary windings have resistance and when any current flows through these windings there must be a drop in potential (voltage).

- The magnetic circuit requires energy to magnetize the circuit first in one direction and then in the other at a rate equal to the alternations of the supply voltage.

- Energy is required to supply the eddy currents which are created in the magnetic circuit.

- Magnetic leakage does exist to some extent in all transformers.

A well-designed transformer deviates from an ideal transformer. For example, an actual distribution transformer rated at 3 kVA with its primary winding connected to a 2400 V supply, and its secondary connected to ten .5 A lamps has a decrease in secondary no-load voltage due to exciting current flowing through the primary. The decrease, due to magnetic leakage, is so small that for all practical purposes it cannot be measured. The no-load primary current is .0625 A. The energy taken from the source by the no-load current (iron loss or core loss) is 30 W. If the primary voltage is maintained constant, the core loss is constant at all loads.

When 50 lamps taking a total of 25 A are connected to the secondary, the secondary voltage (E_s) is 117.36 V. The power loss in the primary coil is 33 W and the power loss in the secondary is 33 W. Therefore, the energy delivered to the lamp is 2934 W (117.36 V × 25 A = 2934 W). The power taken from the source is 3030 W (2934 W + 30 W + 33 W + 33 W = 3030 W). The efficiency of this transformer with a secondary load of 25 A equals 96.83% (2934 W ÷ 3030 W × 100 = 96.83%).

VECTOR DIAGRAMS

The current, voltage, and flux relationships in a transformer are represented in a vector diagram. A *vector* is a quantity involving direction and magnitude.

Ideal Transformer Under No Load

In an ideal transformer in which there is neither resistance nor reactance, reactance (X), in ohms, is equal to the magnetic leakage (or inductance in henrys) multiplied by $2\pi f$ where π equals 3.1416, and where f is the frequency of the circuit. The term $2\pi f$ is frequently expressed as ϖ and inductance in henrys is expressed by the letter L. Then, $2\pi fL = \varpi L$.

Assuming a 1:1 ratio of transformation for simplicity, OE_p represents the impressed primary voltage. See Figure 2-3. Oc represents the primary winding current which produces the flux $O\Phi$, which lags the impressed voltage by 90°. OE_s represents the secondary induced voltage which lags the flux by 90°, therefore 180° behind (in direct opposition to) the impressed primary voltage (E_p).

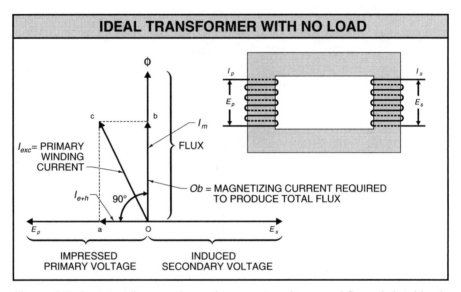

Figure 2-3. A vector diagram shows the current, voltage, and flux relationships in a transformer.

If the magnetic circuit were also an ideal circuit, there would be no iron loss (eddy-current and hysteresis loss) as indicated by the vector Oa and the no-load current would be equal to Ob which is the magnetizing current required to establish the total flux (Φ). Because the magnetic circuit does have eddy-current and hysteresis loss, the total no-load (exciting) current indicated by the vector Oc is the vector sum of Oa and Ob. This no-load current lags the primary impressed voltage (E_p) by the angle E_pOc.

The power factor of the transformer at no load is represented graphically by the cosine of the angle E_pOc or is equal to I_{e+h}, which is the iron loss current divided by IOc, which is the exciting current. Multiplying each of these current values by the primary voltage gives the iron loss in watts and the exciting power. Therefore, the no-load power factor equals $I_{e+h} \div I_{exc}$ which equals the iron loss in watts divided by the exciting watts. Because the iron loss is small compared to the exciting power, the power factor is very low at no load.

Ideal Transformer With Noninductive Load On Secondary

In the vector diagram of an ideal transformer with the secondary having a noninductive load, the lines OE_p, OE_s, $O\Phi$, Oa, Ob, and Oc represent the same quantities as the ideal transformer under no load. See Figure 2-4. In addition, there is a secondary load current (I_s) in phase with the secondary voltage OE_s. I_p represents the additional primary current which must flow in the primary winding, due to the secondary noninductive load.

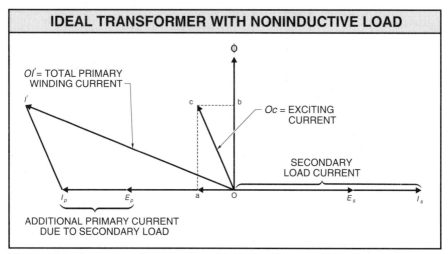

Figure 2-4. The vector diagram of an ideal transformer with a noninductive load on the secondary includes a secondary load current and additional primary current, due to the secondary load.

Assuming a 1:1 ratio of transformation, the primary current (I_p) is equal to the secondary current (I_s) and opposite in direction. The product I_pN_p, which represents the primary ampere-turns, is equal to I_sN_s, which represents the secondary ampere-turns. OI' represents the total primary winding current and is equal to the vectorial sum of the current OI_p (required to balance the secondary current I_s) and the exciting current Oc.

In this diagram, Oc is not drawn to the same scale as OI_p but is shown larger than the actual ratio that exists between these two currents to make the diagram clearer. In actual practice, the exciting current Oc seldom is more than 10% of the load current OI_p. In large transformers, it may be less than 1%. This small value, when added vectorially to OI_p, increases OI_p a very small amount, making OI' (total primary winding current) nearly equal to the load current. The effect of the exciting current with respect to additional loss and heat because of this power loss is disregarded in practice.

However, the noninductive load greatly reduces the angle of lag between the primary impressed voltage (E_p) and the total primary current (I'), greatly improving the power factor of the transformer which is measured by the cosine of the angle between these two vectors.

Ideal Transformer With Highly Inductive Load on Secondary

In the vector diagram of an ideal transformer with the secondary having a highly inductive load, the lines OE_p, OE_s, $O\Phi$, Oa, Ob, and Oc represent the same quantities as an ideal transformer with a noninductive load on the secondary. See Figure 2-5. The secondary load current (I_s) lags the secondary voltage (E_s) by nearly 90° because the load on the secondary is almost a purely inductive load. With such a load, the exciting current becomes more noticeable by making the total primary current (I') greater than the impressed primary current (I_p). The angle E_sOI_s represents the angle of lag of the load current behind the secondary voltage. The cosine of this angle represents the power factor of the secondary load.

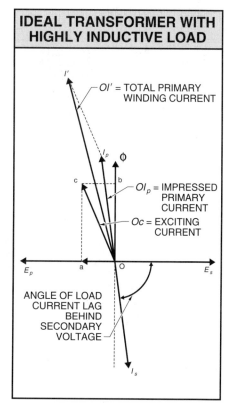

IDEAL TRANSFORMER WITH HIGHLY INDUCTIVE LOAD

I'

OI' = TOTAL PRIMARY WINDING CURRENT

I_p ϕ

OI_p = IMPRESSED PRIMARY CURRENT

Oc = EXCITING CURRENT

E_p a   O E_s

ANGLE OF LOAD CURRENT LAG BEHIND SECONDARY VOLTAGE

I_s

Figure 2-5. In an ideal transformer with the secondary having a highly inductive load, the secondary load current lags the secondary voltage by nearly 90°.

Actual Transformer Vector Diagrams

In an actual transformer carrying a load, the ideal transformer vector diagrams are complicated by resistance and reactance drops in both the primary and secondary windings. To clarify the effects of these two quantities, these values are considered as external to the transformer. See Figure 2-6.

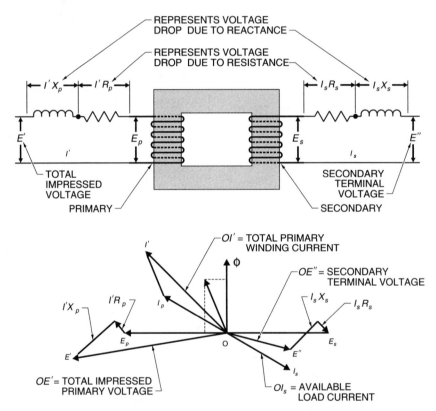

Figure 2-6. The effects of resistance and reactance drops in the windings are considered as external to the transformer to clarify the vector diagram.

X_p, X_s, R_p, and R_s represent the reactance and resistance (in Ω) of the primary and secondary windings. $I'X_p$ and $I'R_p$ represent voltage drops in the primary winding due to reactance and resistance respectively when a load current (I_p) is flowing in the winding. The total impressed voltage (E') applied to the primary winding must overcome the resistance

(R_p) of the primary winding, the reactance (X_p) due to leakage flux, and balance the electromotive force induced in the primary winding by the magnetic flux Φ.

In the secondary winding, I_sX_s and I_sR_s represent the voltage drops due to reactance and resistance when current (I_s) is flowing in the winding. The secondary terminal voltage is E''.

The vector OE', represents the total impressed voltage on the primary. It is the result of combining the resistance drop (vector $I'R_p$), the reactance drop (vector $I'X_p$), and the voltage impressed across the coil (vector OE_p). The resistance drop is in phase with the total current in the primary winding (I'). Thus, vector $I'R_p$ is parallel to OI'. The reactance drop $I'X_p$ is 90° out-of-phase with I', and the vector $I'X_p$ is at a right angle to OI'.

Assuming a 1:1 ratio of transformation, the total electromotive force (E_s) induced in the secondary coil is equal and opposite to OE_p. If the turns ratio of primary to secondary is N_p/N_s, the induced voltage OE_s equals $N_s/N_p \times OE_p$. A portion of this secondary induced voltage is used to overcome the resistance and reactance of the secondary winding. The remainder of the voltage is available at the terminals of the secondary winding for sending the load current (I_s) through the external circuit.

The secondary resistance drop is represented by I_sR_s and the reactance drop is represented by I_sX_s. The remainder of the voltage, which represents the secondary terminal voltage when the transformer is supplying a load of I_s amps, is represented by the line OE''.

The ratio of total impressed voltage on the primary (E') to secondary terminal voltage (E'') is no longer 1:1, and these voltages are no longer in direct opposition to each other. This ratio and the phase angle between E' and E'' depend on the relative values of IR and IX in the primary and secondary windings. The relative values of IR and IX depend on the resistance and reactance of the transformer and the current in the windings. The resistance and reactance values for any given transformer are constant and can be measured. The current is a function of the impedance of the load.

Voltage Ratio. It is common practice in distribution and power transformers to consider the voltage ratio equal to the turns ratio of the primary and secondary windings, even with all the variables affecting the voltage ratio. While this error is considered negligible, it is an error which must be taken into account for measuring devices such as voltmeters, ammeters, wattmeters, and varmeters. This error is compensated for by slightly changing the turns ratio.

The no-load (exciting current) is the only factor that affects the ideal relationship between primary and secondary currents. However, the relationship between primary and secondary electromotive forces are affected by the resistance of both primary and secondary windings and by the total magnetic leakage (reactance) of these windings. With a noninductive load, only the resistance affects this ratio, and the secondary terminal voltage decreases with an increase in load. This is due almost entirely to the *IR* drop of the primary and secondary coils. With an inductive load, the resistance drop is less perceptible and the reactance becomes a major factor in lowering the secondary voltage.

If the secondary load is highly inductive, the load current (I_s) and the total primary current (I') are similar to the values of an ideal transformer. The I_sR_s and $I'R_p$ drops are in phase with these respective currents and nearly at right angles to the voltages E_s and E_p. Therefore, when they are added and subtracted respectively from E_s and E_p, they have little effect.

However, the $I'X_p$ and I_sX_s drops, which are at right angles to I' and I_s respectively, are nearly parallel to E_p and E_s and therefore add and subtract directly from E_p and E_s respectively by an amount equal to their arithmetical value.

It is impossible to measure the magnetic leakage of the primary and secondary separately. It is common practice to refer to the total magnetic leakage in terms of either the primary or secondary winding. If the reactance is expressed in ohms with respect to the primary winding, it is expressed in ohms with respect to the secondary winding by multiplying this value in ohms by the square of the ratio of secondary to primary turns. Conversely, if the reactance is expressed in ohms with respect to the secondary winding, it is expressed in ohms with respect to the primary winding by multiplying the reactance by the square of the ratio of primary to secondary turns.

For example, the primary of a 10 kVA transformer having a primary voltage of 1000 V and a secondary voltage of 100 V (no-load turns ratio of 1000:100 or 10:1), has a primary resistance of 1.5 Ω and a secondary coil resistance of .015 Ω. The leakage reactance (ϖL) of the secondary with respect to the primary winding is 5 Ω.

The secondary coil delivers 100 A to a noninductive load and, neglecting the no-load losses, the primary takes 10 A from the 1000 V supply. The voltage drop of electromotive force in the primary coil is 15 V (10 V × 1.5 V = 15 V).

The current I_s and I_p, neglecting no-load losses, is parallel to E_s and E_p respectively because the load is noninductive. Therefore, the 15 V drop is parallel to E_p and must be added directly to E_p to obtain the impressed voltage E'. The primary induced voltage (E_p) equals $E' - I'R_p$. See Figure 2-7. In this example, the primary induced voltage equals 985 V (1000 V − 15 V = 985 V). The voltage drop, due to the leakage reactance of 5 Ω represented by $I'X_p$, is equal to 50 V (10 V × 5 V = 50 V). However, this 50 V is almost at right angles to E'. Therefore, when subtracted vectorially it does not appreciably lower the voltage value. The total electromotive force induced in the secondary coil equals 98.5 V (100 V ÷ 1000 V × 985 V = 98.5 V).

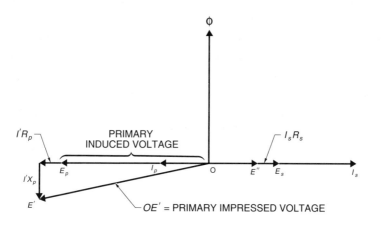

Figure 2-7. Neglecting no-load losses, the total primary induced voltage of a transformer with a noninductive load equals the primary impressed voltage less the drop of electromotive force in the primary coil.

The voltage drop of electromotive force due to resistance in the secondary coil is equal to 1.5 V (100 V × .015 V = 1.5 V). This is parallel to E_s because of the noninductive load. Therefore, it subtracts directly from the secondary induced voltage E_s. The secondary terminal voltage equals 97 V (98.5 V − 1.5 V = 97 V).

If the secondary delivered the 100 A of current to a highly inductive circuit, the 100 A would be nearly 90° behind E_s, and the primary current of 10 A represented by I_p would be nearly 90° behind E_p in phase relation. Therefore, the drop, due to the leakage reactance which is equal to 50 V, is nearly parallel to E_p so that E_p is nearly equal to 950 V (1000 V − 50 V = 950 V). Because the current drops of the primary and secondary windings equal 15 V and 1.5 V respectively and are in phase with their respective currents (I_p and I_s), they must be nearly 90° from the voltages E_p and E_s. Therefore, they have no appreciable effect in lessening these voltages.

Because the total leakage reactance was expressed as 5 Ω when referring to the primary winding, the I_sX_s drop does not exist. Therefore, the secondary terminal voltage E'' equals nearly 95 V ($^{100}/_{1000} \times 950$ V = 95 V). Expressing the leakage reactance in terms of the secondary winding gives .05 Ω [$(^{100}/_{1000})^2 \times 5$ Ω = .05 Ω].

The drop, due to this leakage reactance with 100 A flowing, equals 5 V (.05 Ω × 100 A = 5 V). Because the current drops are negligible in both windings, the secondary induced voltage equals 100 V ($^{100}/_{1000} \times 1000$ V = 100 V). The secondary terminal voltage equals the secondary induced voltage less the drop due to leakage reactance or 95 V (100 V − 5 V = 95 V).

The method used in measuring this leakage reactance determines whether it is expressed in terms of the primary or secondary winding. It is most commonly expressed in terms of the high-voltage winding which is the primary winding in a step-down transformer and the secondary winding in a step-up transformer.

3 TRANSFORMER CLASSIFICATION

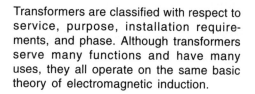

Transformers are classified with respect to service, purpose, installation requirements, and phase. Although transformers serve many functions and have many uses, they all operate on the same basic theory of electromagnetic induction.

TRANSFORMER CLASSIFICATION

A *transformer* is a device that raises or lowers the voltage of alternating current of the original source. Transformers are classified with respect to service, purpose, installation requirements, and phase. They are normally referred to by the purpose and service they perform.

SERVICE CLASSIFICATION

Transformers are classified with respect to service into six general categories. These categories include voltage transformation, voltage regulation, current regulation, metering and protection, accessory, and communication. Each general category may be subclassified into a number of types, such as instrument, current, potential, constant-current, series transformers for street lighting, small power, control and signal, electric sign, chime and bell ringing, neon sign, generator step-up, substation, mobile, distribution, grounding, isolation, filtering, radio, TV, and audio transformers.

These groups are not sharply defined. There are similarities in the design and construction of all of these groups. For example, the dividing line between distribution and substation transformers is more arbitrary than fundamental.

51

Instrument Transformers

Instrument transformers are transformers that step down the voltage or current of a circuit to a low value that can be effectively and safely used for the operation of instruments. These include ammeters, voltmeters, wattmeters, varmeters, relays used for various protective purposes, telemetering used for indications at remote areas, and dispatching energy. Instrument transformers also insulate the instrument, relay, or telemetering equipment from the voltage of the power circuit.

Instrument transformers, including those for either current or voltage transformation, have a small volt-ampere capacity, which is necessary to provide the energy required by the measuring instrument with which they are used. The requirement is for the utmost accuracy of voltage and current transformation, particularly when the transformers are used in connection with the metering of power. The errors that must be compensated for are those in the ratio of transformation and those pertaining to the relative phase position of impressed and delivered electromotive forces and currents.

Although instrument transformers function on the same basic principle as power transformers, their construction may be quite different. Low-voltage instrument transformers, particularly those for switchboard applications are operated without oil. High-voltage instrument transformers are immersed in oil for insulation purposes. While most instrument transformers are small in size, the high-voltage units are very large compared to the devices to which they are connected. When primary transmission systems are tied together, transformers of very high voltage are used for measuring the interchange of power from one system to another.

Current Transformers

A *current transformer (series transformer)* is a transformer that has the primary, normally consisting of one or more turns, connected in series with the line. When the primary has a large current rating, the primary winding may consist of a straight conductor passing through the current of the magnetic circuit. See Figure 3-1. This single conductor may be a portion of the current-carrying bus or any conductor that must be measured or protected.

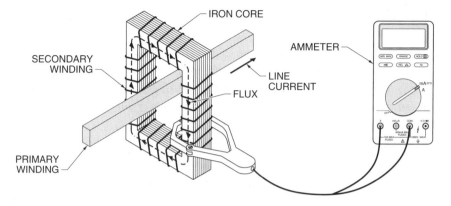

Figure 3-1. A current transformer has the primary connected in series with the line.

The secondary winding, consisting of many turns of insulated wire, is wound around an iron core. The secondary of a current transformer is designed to produce 5 A when the rated current is flowing in the primary. The secondary of a current transformer is always rated at 5 A regardless of the amperage rating of the primary. See Figure 3-2. This enables the production of standardized current devices which are rated at 5 A. The ratio of current transformation is approximately the inverse ratio of the number of turns.

ABB Power T&D Company Inc.

Figure 3-2. The secondary of a current transformer is designed to produce 5 A when the rated current is flowing in the primary.

The current rating of the primary winding of a current transformer is determined by the maximum value of the load current to be measured. For example, if the current rating is 400 A and the secondary winding has a rating of 5 A, the ratio between the primary and secondary is 400:5 or 80:1. This means that the secondary winding has 80 times as many turns as the primary. If the primary winding has 5 turns, the secondary winding has 400 turns.

The ratio of the primary to the secondary current is inversely proportional to the ratio of primary to secondary turns. For example, a current transformer with a ratio of 400:5 having a load current of 400 A has a secondary current of 5 A. The same transformer with a load current of 300 A has a secondary current of 3.75 A. The same transformer with a load current of 100 A has a secondary current of 1.25 A.

The required ratio of a current transformer depends on the maximum line current it is expected to carry. Ratings may vary from 5:5 to 8000:5. This assumes an ideal transformer in which there is neither resistance nor reactance in the primary and secondary windings, no energy is required to energize the core, and the load on the secondary, usually the current coil of an ammeter, has no impedance. Because these conditions actually exist in the current transformer circuit, they must be compensated for, otherwise the errors introduced become of great importance and may result in a considerable amount of power being unaccounted for. This is especially true when metering large quantities of power with an integrating meter whose current coil is connected to the secondary of the instrument transformer.

These factors, which introduce errors, are partially compensated for in the design of instrument transformers by using a magnetic circuit consisting of very high permeability steel at the flux density at which the transformer operates. The flux density is kept at a low value so that the disturbing element of magnetizing current is low. Additional compensation is obtained by decreasing the number of turns of the secondary below that determined by the inverse ratio of currents in the primary and secondary. This allows a greater current flow in the secondary and establishes the desired current ratio at the particular secondary load for which greatest accuracy is desired.

Unlike a voltage transformer, the secondary current of an instrument transformer is not determined by the impedance of the load connected across the secondary. The secondary current depends entirely on the amount of current flowing through the primary. If the secondary circuit is open, all the current flowing through the primary winding must be magnetizing current. Because the magnetic circuit is designed for a low magnetizing current when the transformer is loaded, this large increase in magnetizing current builds up an enormous flux in the magnetic circuit. This causes the transformer to act as a step-up transformer, inducing an excessively high voltage across the terminals of the secondary. Therefore, a current transformer should always have its secondary shorted when not connected to an external load.

A manually- or automatically-operated secondary short-circuit device is supplied as a standard accessory by most manufacturers of new current transformers. See Figure 3-3. However, extreme care must be taken to ensure that before working on a current circuit, the circuit has been jumped and the series path is reestablished ahead of the point being worked on. The three types of current transformers in general use are the wound, window, and bar.

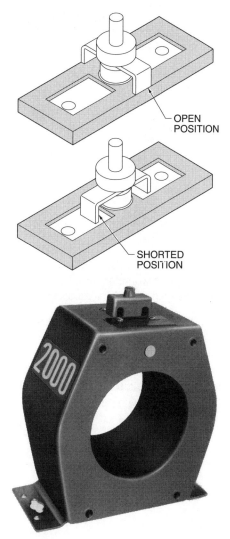

ABB Power T&D Company Inc.

Figure 3-3. A manually- or automatically-operated secondary short-circuit device is supplied as a standard accessory by most manufacturers of new current transformers.

Wound. *Wound-current transformers* are transformers that have separate primary and secondary windings mounted on a laminated core. See Figure 3-4. A wound-current transformer is designed so that its primary winding consists of one or more turns of large cross-section wire connected in series with the circuit to be measured.

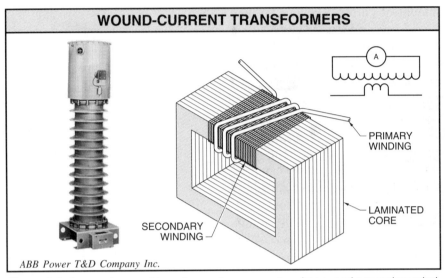

WOUND-CURRENT TRANSFORMERS

ABB Power T&D Company Inc.

Figure 3-4. A wound-current transformer has separate primary and secondary windings mounted on a laminated core.

Wound-current transformers may be constructed so they have a dual-ratio. This is done by having two primary windings which may be connected in series or parallel by links. See Figure 3-5. The external primary leads are connected to terminals 2 and 3. For the highest ratio, the windings are connected in parallel by connecting terminal 1 to 2 and 3 to 4.

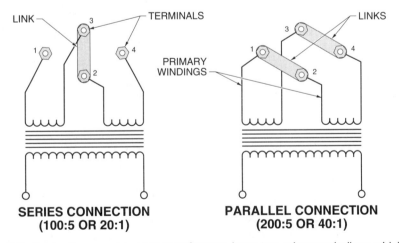

SERIES CONNECTION
(100:5 OR 20:1)

PARALLEL CONNECTION
(200:5 OR 40:1)

Figure 3-5. Dual-ratio wound-current transformers have two primary windings which may be connected in series or parallel by links.

Window. A *window-type current transformer* consists of a cylindrical-ring core of thin iron lamination similar to a stack of washers. See Figure 3-6. Copper wire is wound around the core which forms the secondary winding. Taps are brought out from the winding.

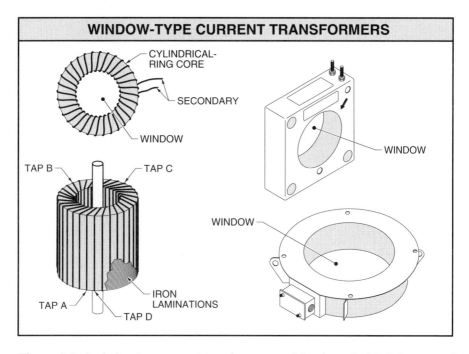

WINDOW-TYPE CURRENT TRANSFORMERS

Figure 3-6. A window-type current transformer consists of a cylindrical-ring core of thin iron laminations.

Window-type current transformers are mounted in the bushing of an oil-circuit breaker. See Figure 3-7. The high-voltage conductor, which is the lead to the terminal on the breaker bushing, forms the one primary turn of the transformer. A window-type current transformer is less accurate at low ratings than a wound-current transformer.

The operating ratio of window-type current transformers may be changed by varying the number of wires passing through the window of the transformer. The ratio is not affected by the direction of wires in the window. The direction of the current in the conductors does affect the metering. To calculate the operating ratio of a window-type current transformer, apply the formula:

$$O_r = N_r \times \frac{1}{w}$$

where

O_r = operating ratio

N_r = nameplate ratio

1 = constant

w = number of wires in window

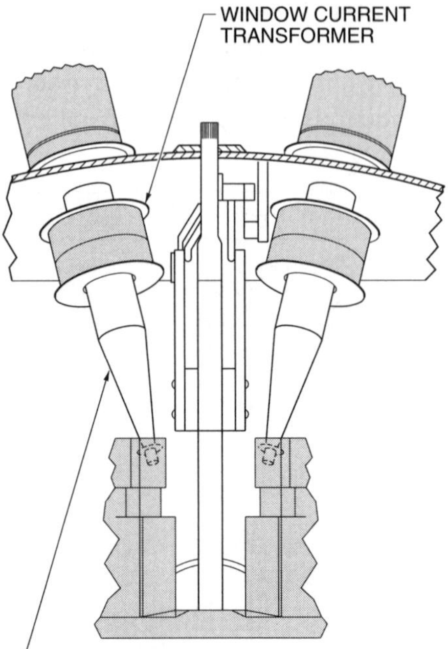

Example: Calculating Operating Ratio

What is the operating ratio of a 300:5 window-type current transformer that has one conductor through its window?

$$O_r = N_r \times \frac{1}{w}$$

$$O_r = \frac{300}{5} \times \frac{1}{1}$$

$$O_r = \frac{300}{5}$$

$$O_r = \mathbf{60{:}1}$$

WINDOW CURRENT TRANSFORMER

HIGH-VOLTAGE CONDUCTOR

Figure 3-7. Window-type current transformers are mounted in the bushing of an oil-circuit breaker.

If two conductors are passed through the same 300:5 current transformer, the operating ratio is 30:1 ($\frac{300}{5} \times \frac{1}{2} = 30{:}1$). Therefore, with 150 A in the conductor, there is 5 A in the secondary of the connection.

A window-type current transformer has subtractive polarity, as does any other instrument transformer. The primary and secondary polarity is indicated by a mark on the transformer. If the primary conductor enters the window from the end of the current transformer marked with the polarity mark, and the current in this conductor goes in this same direction, the secondary current leaves the terminal of the window-type current transformer marked with the polarity mark.

A special connection which uses two conductors, one from each phase, is used to make a three-wire current transformer. See Figure 3-8. Each conductor must pass through the current transformer in an opposite direction to ensure proper registration for this type of metering. To determine the secondary current, the current in each phase is multiplied by the operating ratio of the transformer.

Example: Calculating Secondary Current

What is the secondary current of a 50 A per phase load on a current transformer that has a rating of 200:5 and two wires in the window?

1. Calculate operating ratio.

$$O_r = N_r \times \frac{1}{w}$$

$$O_r = \frac{200}{5} \times \frac{1}{2}$$

$$O_r = \frac{40}{1} \times \frac{1}{2}$$

$$O_r = 20{:}1$$

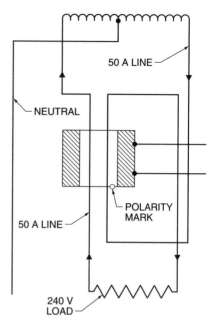

Figure 3-8. A special connection using two conductors is used to make a three-wire current transformer.

2. Multiply operating ratio by current per phase.

$$I_s = \frac{50 \times 1}{20}$$

$$I_s = \textbf{2.5 A}$$

For 50 A in each phase, the secondary current is 2.5 A. If the current is 100 A, there is 5 A in the meter coil and secondary of the current transformer.

Two standard two-wire current transformers are used to develop a three-wire, 3ϕ connection. See Figure 3-9. The current transformers used must have the same nameplate ratio and the same number of turns through the window. Also, the direction of the phase conductor through the window of each current transformer must be the same.

For example, a 240 V, 3ϕ, 100 A motor load with no lighting load is connected to two current transformers. Both current transformers have the same nameplate ratio (400:5) and have two turns through the window (of the same phase). That is, phase A current transformer has two turns of conductor A and no other phase conductor is involved. The operating ratio for both current transformers is 40:1 ($^{400}/_5 \times ^1/_2$ = 40:1). The secondary current in the meter coil is 2.5 A ($^{100} \times ^1/_{40}$ = 2.5 A) in each current transformer if the motor load draws 100 A.

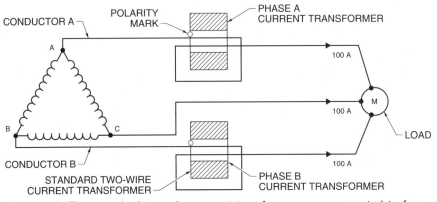

Figure 3-9. Two standard two-wire current transformers are connected to form a three-wire, 3ϕ connection.

Two transformers may be connected to form a four-wire, 3ϕ delta current transformer connection. One current transformer is used as a three-wire current transformer and one as a two-wire current transformer. See Figure 3-10.

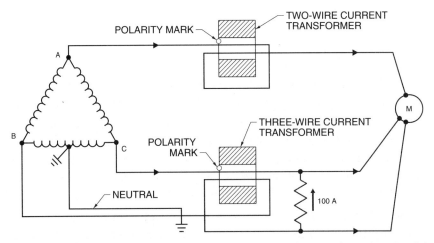

Figure 3-10. Two transformers may be connected to form a four-wire, 3ϕ delta current transformer connection.

The nameplate ratio of both current transformers is the same. To avoid ratio errors, it is necessary to have the same number of turns in the window of each current transformer. Also, the direction of the conductor through the window must be the same.

For example, the operating ratio of a 120/240 V, 3ϕ, 400 A, four-wire balanced delta load and a 240 V, 1ϕ, 100 A load without a 120 V lighting load is 80:1 ($^{800}/_5 \times \frac{1}{2} = 80:1$). Both current transformers have two turns. In one case, both turns are from the same phase (two-wire current transformer), and in the other case, there is one turn from two phases (three-wire current transformer).

For the 240 V, 3ϕ motor load, the current transformer secondary current is 5 A ($^{400 \times}/_{80} = 5$ A) in both the three-wire and two-wire current transformers. In the 240 V, 1ϕ load, the current transformer secondary current in the three-wire transformer is 1.25 A ($^{100 \times}/_{80} = 1.25$ A). In the three-wire current transformer, the secondary current is the vector sum of the 1ϕ current of 1.25 A and the 3ϕ current of 5 A. The magnitude depends on the power factor of each load. The current flows through one current coil of a two element, 3ϕ, three-wire meter, while the 5 A output of the two-wire current transformer flows through the other current coil of the meter.

Bar. A *bar current transformer* is a window current transformer with a solid bar placed permanently through the window. See Figure 3-11. A bar current transformer can withstand the stresses of heavy overcurrent, and the tendency to assume a circular shape is eliminated. To avoid magnetic stresses that could destroy the bus and damage the transformer, care must be taken to properly mount these transformers with respect to adjacent conductors.

Current Transformer Classes. Current transformers may be classified as dry, oil-, or compound-filled for indoor or outdoor use. The application determines the type used. Normally the dry or compound-filled type is used for voltages below 22,000 V and either the compound- or oil-filled is used for voltages above 22,000 V. As the voltage increases, insulation becomes more important. The amount and type of insulation depends on the voltage at which the transformer is used. As the voltage increases, the primary winding has to be more highly insulated from the secondary.

Polarity Markings. Polarity is required when current transformers are used with meters or relays which rely on magnitude and phase position.

Polarity is indicated by a mark on one primary and one secondary terminal (normally a white dot). When current is flowing toward the marked primary terminal, it is flowing away from the marked secondary terminal. This marking corresponds to the H1 and X1 terminals on power transformers. All standard instrument transformers are subtractive polarity.

BAR CURRENT TRANSFORMERS

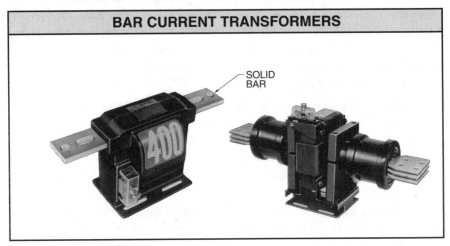

ABB Power T&D Company Inc.

Figure 3-11. A bar current transformer is a window current transformer with a solid bar placed permanently through the window.

Current Transformer Precautions. Current transformers differ from potential and other transformers in that the primary winding is designed for connection in series with the line current at all times. The secondary circuit of a current transformer should never be open-circuited when there is current flowing in the primary winding. Under open-circuit conditions, the primary current becomes an exciting current which causes a high voltage to be induced in the secondary winding. This voltage may be sufficient to puncture the insulation and may produce a dangerous high-voltage shock to anyone who comes in contact with the open circuit secondary.

Anyone working with current transformers must always ensure the secondary winding circuit of a current transformer is closed or shorted. When in service, all current transformers should have their secondary winding grounded because an electrostatic field builds up in the secondary winding.

Current Transformer Accuracy. A current transformer differs from an ordinary transformer in that its primary current is determined entirely by the load on the system and not by its own secondary load. A current transformer must change the magnitude only if the current being measured changes. The accuracy of a current transformer must be known so that any errors can be included in the computation of the overall measurement, or the errors must be within the limits of a specified small value so that they may be disregarded as insignificant.

The design and construction of a current transformer, the circuit's current and frequency, and the load imposed on the secondary circuit of the transformer all affect its accuracy. In general, the greater the load, the greater the error. The secondary current is proportionately less than the primary current because a portion of the primary current is required to magnetize the core and supply the core losses. These small errors are classified as ratio and phase-angle errors.

Current transformers may be given standard accuracy class ratings which define the performance at overcurrent values for relaying purposes. These ratings are based on the standard secondary terminal voltage a transformer delivers without exceeding a standard percent ratio error. This classification appears as 2.5% (H or L) 200, 5% (H or L) 200, or 10% (H or L) 200. The first term is the maximum percent ratio error. This may be either 2.5%, 5%, or 10%. The second term may be H or L. Current transformers fall into two classes. The first class, designated by the letter H, has a nearly constant percentage error when delivering a fixed secondary voltage over a wide range of secondary current. The second class, designated by the letter L, has a nearly constant magnitude error under the same conditions. In general, H class transformers are the wound-type, and L class transformers are the window-type. The third term is the secondary voltage which can be delivered at 20 times the rated secondary current without exceeding the ratio error. The secondary voltage class may be 10 V, 20 V, 50 V, 100 V, 200 V, 400 V, or 800 V.

Potential (Voltage) Transformers

A potential (voltage) transformer operates on the same principle as a power or distribution transformer. The main difference is that the capacity of a potential transformer is relatively small as compared with power transformers. Potential transformers have ratings of 100 VA to 500 VA. The low-voltage side is normally wound for 120 V. See Figure 3-12.

POTENTIAL (VOLTAGE) TRANSFORMERS

OUTDOOR 7200 V

FUSES

INDOOR 2400 V, FUSED

OUTDOOR 2400 V

INDOOR AND OUTDOOR 600 V

ABB Power T&D Company Inc.

Figure 3-12. The capacity of potential transformers is relatively small as compared with power transformers.

The load on the low-voltage side consists of the potential coils of various instruments. In some cases, potential coils of relays and other control equipment are also connected to the secondary of the potential transformer. In most cases, the load is relatively light, and it is not necessary to have a capacity of potential transformers greater than 500 VA.

The primary windings of a potential transformer are designed to be connected in parallel with the circuit in question. Because the secondary is normally 120 V, standard instruments and relays can be used, and the worker is protected against dangerous high voltage. Potential transformers may also be used for isolation purposes. They are sometimes used to provide a potential of a polarity reversed in relation to the potential of the power circuit. The ratio range of a potential transformer is from 1:1 to 345 kV to 120 V.

Instrument loads are normally connected through potential and current transformers to a high-voltage 1ϕ line. See Figure 3-13. The loads on the transformers include an ammeter, wattmeter, watt-hour meter, and voltmeter.

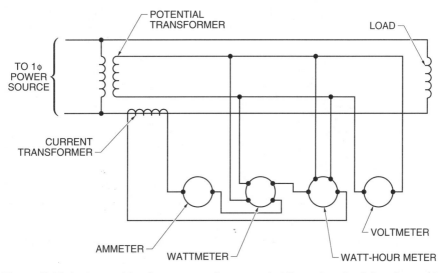

Figure 3-13. Instrument loads are normally connected through potential and current transformers to a high-voltage 1ϕ line.

Potential Transformer Types. Care must be taken in choosing a potential transformer which is suitable for the conditions to be encountered in applications where high voltages are encountered. A potential transformer designed for outdoor service is normally used on circuits greater than 25,000 V. This may be a single-bushing or double-bushing type. On circuits of 25,000 V and lower, either indoor or outdoor transformers are employed. Single-bushing potential transformers are used where phase-to-ground voltages are measured. See Figure 3-14.

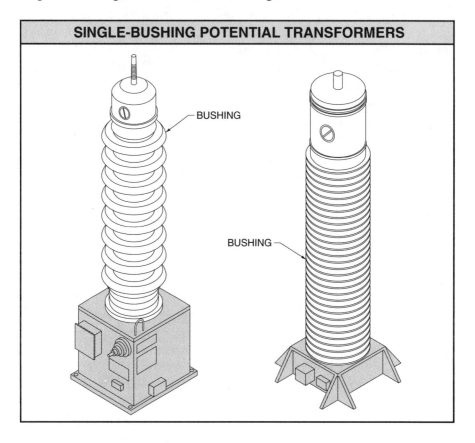

SINGLE-BUSHING POTENTIAL TRANSFORMERS

BUSHING

BUSHING

Figure 3-14. Single-bushing potential transformers are used where phase-to-ground voltages are measured.

A double-bushing potential transformer has two bushings on the high-voltage winding that are capable of being connected to a circuit. The voltage of the bushings is the same as that on the nameplate of the transformer. The bushings may be connected either phase-to-phase or phase-to-neutral. On a 3ϕ system, a double-bushing potential transformer may be connected delta, wye, or open-delta. See Figure 3-15.

DOUBLE-BUSHING POTENTIAL TRANSFORMERS

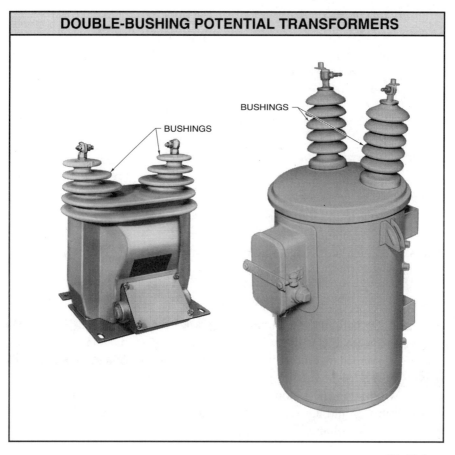

BUSHINGS

BUSHINGS

BUSHINGS

Allis Chalmers

Figure 3-15. A double-bushing potential transformer has two bushings on the high-voltage winding that are capable of being connected to a circuit.

Capacitor voltage transformers are generally used on systems having a voltage that exceeds 100 kV because capacitors are easier to design than magnetic transformers for use at high voltages. See Figure 3-16. A capacitor voltage transformer consists of a capacitive potential divider and a potential transformer. By stacking a number of capacitors in series according to the desired voltage, the advantages of standardization and lower manufacturing cost are obtained. Each transformer is provided with taps to accurately adjust the output according to phase and magnitude. The spark gaps protect both the bushing and secondary instruments or relays.

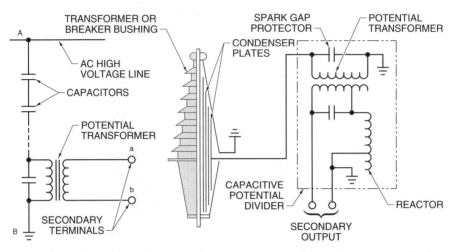

Figure 3-16. Capacitor voltage transformers are generally used on systems having a voltage that exceeds 100 kV.

Potential Transformer Accuracy. The two causes of errors in potential transformers are the exciting current, necessary to magnetize the core, causing an impedance drop in the primary winding, and the load current drawn by the load, causing an impedance drop in the primary and secondary windings. The secondary voltage is less than the primary to secondary ratio would indicate and the phase angle between the primary and secondary voltage is reversed.

Constant-Current Transformers

A constant-current transformer is used extensively to supply a constant current for series street lighting that uses incandescent or gas-discharge lamps connected in series. A *constant-current source* is a voltage source in which the voltage impressed on the secondary series circuit is automatically increased or decreased as the current decreases or increases, keeping the current of the secondary circuit at a constant value. The usual method of feeding a constant-current circuit is by the use of a constant-current moving-coil regulator. See Figure 3-17. The primary coil is energized by a constant voltage source.

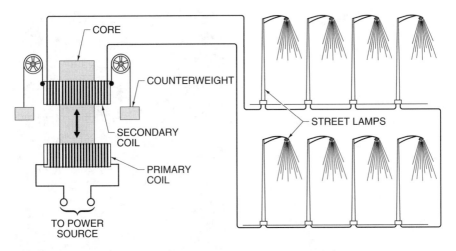

Figure 3-17. A constant-current circuit is normally supplied power by a constant-current moving-coil regulator.

A constant-current transformer is built for either air or oil cooling. Oil is often used because it is an excellent agent for insulation, cooling, and lubrication. Constant-current transformers are constructed for outdoor pole-mounting, subway mounting, and indoor station use. Indoor station and subway mounting types are built for indoor use and normally serve from 15 to more than 200 lamps. The outdoor pole-mounting type is normally built in smaller sizes and is normally installed close to the area served by the street-lighting circuit. See Figure 3-18.

The outdoor pole-mounting type is weatherproof and is normally oil insulated. The primary coil of a constant-current transformer is normally wound for 2400 V but can be wound for any reasonable higher voltage up to about 10,000 V. The secondary is wound for the voltage required for operating the number of lamps in the series circuit. The constant current in the secondary circuit is normally 6.6 A. The standard ratings of an outdoor pole-mounting type transformer are 5 kVA, 10 kVA, 15 kVA, 20 kVA, 25 kVA, 30 kVA, 35 kVA, 40 kVA, 50 kVA, 60 kVA, and 70 kVA.

The efficiency of a constant-current transformer is about 96% at full load for a 100 lamp transformer. The power factor, which depends on the magnetic leakage, is low at all loads, reaching 75% to 80% at full load and decreasing almost proportionally for lower loads.

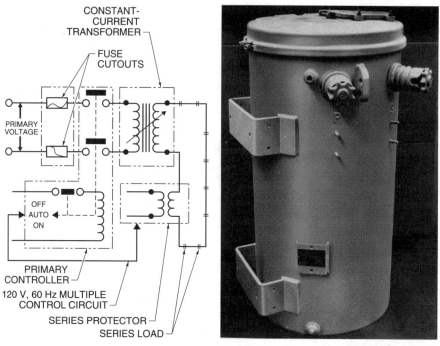

CONSTANT-
CURRENT
TRANSFORMER

FUSE
CUTOUTS

PRIMARY
VOLTAGE

OFF
AUTO
ON

PRIMARY
CONTROLLER

120 V, 60 Hz MULTIPLE
CONTROL CIRCUIT

SERIES PROTECTOR

SERIES LOAD

General Electric Co.

Figure 3-18. Constant-current transformers constructed for outdoor pole-mounting are normally built in small sizes and installed close to the area served by the street-lighting circuit.

a series circuit burns out, the circuit must be reestablished in some manner or all of the lamps in the circuit do not burn. The circuit may be reestablished by the use of a film cutout or it may be maintained by a small reactance coil shunted around the lamp. See Figure 3-19.

A *film cutout* is a device in the base of a lamp that punctures and reestablishes the circuit when subjected to a voltage above a certain critical value. The device consists of two metal contacts separated by a thin film of insulating material. When the lamp burns out, the entire voltage of the circuit is momentarily impressed across the contacts of the device, puncturing the insulation between the metal contacts and shorting out the lamp. This reestablishes the continuity of the series circuit.

Regulation is sometimes employed to maintain the voltage at a given level. This is normally done without the use of a moving-coil regulator. In this case, a constant-voltage transformer is connected in series with a reactance coil. The reactance coil and the transformer have taps by which the voltage of the circuit, as well as the amount of reactance, can be adjusted to the number of lamps in the circuit. In the case of failure of individual lamps, the circuit is reestablished by film cutouts. The reactance coil is used to produce a circuit relatively high in reactance. A change in the resistance of the circuit, by adding or taking away a lamp or two, does not materially change the flow of current because the impedance of the circuit is largely made up of reactance. An objection to series circuits is the low-power factor resulting from the reactance coil of the reactance coil/voltage transformer combination or from the leakage reactance of the moving-coil regulator.

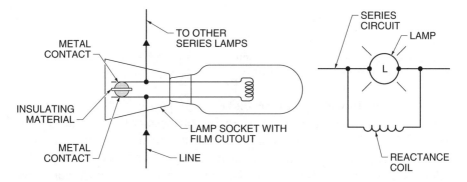

Figure 3-19. A series circuit may be reestablished when a lamp burns out by the use of a film cutout or it may be maintained by a small reactance coil shunted around the lamp.

Constant-Current Regulator Operation. A constant-current regulator is constructed so that the primary and secondary coil can move with respect to each other. Either the primary or the secondary coil is movable. The secondary coil is suspended from a lever which is counterweighted. Constant-current regulators are normally equipped with a dashpot to prevent rapid changes or any hunting action of the movable coils. See Figure 3-20.

When the secondary coil is free to move up or down and delivers a certain amount of current to the lamps on the secondary side, the magnetic flux (Φ) which passes through the primary coil, flows partly through the secondary coil as the useful flux (U), and partly leaks between the primary and secondary coils as the leakage flux (L). The magnetic circuit is such that the coils are mounted on a central leg of the core and the useful flux has two parallel paths through which it completes its circuit. The two parallel paths are the upper and lower yoke sections and the outer legs.

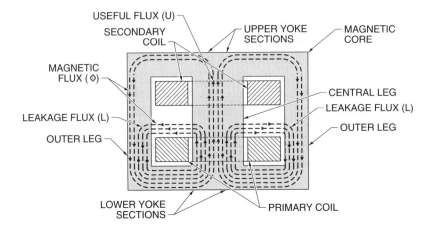

Figure 3-20. A moving-coil regulator is constructed so that the primary and secondary coil can move with respect to each other.

The current flowing in the secondary is opposite in direction to that of the primary. There is a repulsion between these two coils, due to the leakage flux of both coils producing a strong magnetic field in the air space between the two coils. The counterweight is adjusted so it nearly balances the weight of the coil. The remainder of the coil weight is balanced by the upward push of the leakage flux produced with a given amount of current flowing through the secondary coil.

The load resistance decreases if the secondary load decreases (by short-circuiting one or more lamps). The secondary current increases because of the decreased load resistance. This increases the primary current which increases the leakage flux and produces an increased repelling force between the two coils. The secondary coil, due to the unbalanced action of the counterweight, moves away from the primary coil at the same time the useful flux is decreased by the action of the increased secondary current. The induced electromotive force in the secondary decreases, and the upward movement of the secondary coil continues until the induced electromotive force and the secondary current are restored to their normal value.

Similarly, an increase of resistance in the secondary circuit (produced by the addition of one or more lamps) causes a momentary decrease of secondary current. This decreases the upward force on the secondary coil. The coil moves downward until the secondary current is increased to its normal value. Within working limits, the magnetic repulsion between the fixed and moving coils of the regulator, for a given position of coils, is proportional to the current flowing in the coils. This makes the transformer capable of adjustment so that any desired current may be maintained by changing the amount of counterweight.

The secondary current in a series street-lighting system is normally 6.6 A. The action between the primary and secondary coils is such that this value is maintained nearly constant over the entire working range of the regulator.

Series Transformers

Series transformers operate on the same principle as current transformers. The primary winding is connected in series with the main series circuit. The secondary winding is used to feed series incandescent or mercury-vapor street lamps. They are designed to operate at a different current than that of the main series circuit. They may isolate a lamp which is designed to operate at the same current but at a much lower voltage than that of the main series circuit.

Series transformers are manufactured with one or more secondaries to feed one or more series lamps from each secondary. The normal current rating of the secondary of a series transformer is 6.6 A, 15 A, or 20 A. These series transformers are made with ratings of .25 kW, .5 kW, 1 kW, 2 kW, 3 kW, 4 kW, 5 kW, 6 kW, 7.5 kW, and 9 kW. See Figure 3-21.

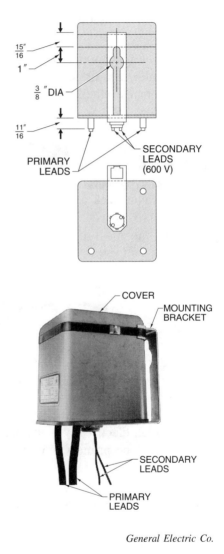

General Electric Co.

Figure 3-21. A series ballast and transformer are used to supply a mercury-vapor lamps from series street-lighting circuits.

A series ballast and transformer is used to supply a mercury-vapor lamp from a series street-lighting circuit. This type of transformer is used to supply a single lamp from a series 6.6 A or 20 A main circuit.

Small Power Transformers

Small power transformers are constant-potential, self-air-cooled transformers that are available in standard sizes of 75 W, 150 W, 225 W, and 300 W. A common use of a small power transformer is to step down the supply voltage to lamps which serve localized lighting at a machine to a safe value, such as 6 V, 32 V, 64 V, or 115 V when the supply circuit to the machine is a relatively high voltage, such as 460 V, 3φ. See Figure 3-22.

Another use of small power transformers is to serve underwater lighting systems in swimming pools. This type of transformer is available for weatherproof outdoor service in various voltage and kVA ratings. This special transformer is a two-winding insulated trans-former which electrically isolates the secondary swimming pool lighting system from the primary supply system.

General Electric Co.

Figure 3-22. Small power transformers are used to step down the supply voltage to lamps which serve localized lighting.

Control and Signal Transformers

Control and signal transformers are self-air-cooled, constant-potential transformers which are used to step down the voltage to supply signal circuits or control circuits of electrically-operated switches. See Figure 3-23. A common use of control and signal transformers is to step down the motor control circuit voltage to a safe value of 120 V when the motor and its controller are supplied from a 460 V, 3ϕ power circuit. Small control transformers operating at 120/25 V are commonplace in residential buildings, supplying control voltages for heating and air conditioning.

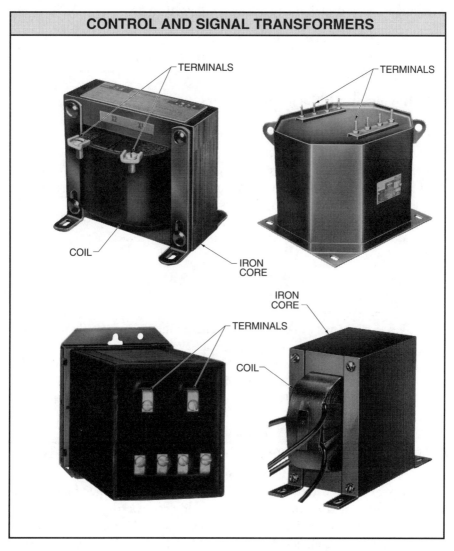

ABB Power T&D Company Inc.

Figure 3-23. Control and signal transformers are used to step down the voltage to supply signal circuits or control circuits of electrically-operated switches.

Electric Sign Transformers

Electric sign transformers are constant-potential transformers that are used to step down a 120/208 V or 115/230 V circuit to a lower voltage to serve low-voltage incandescent lamps in signs. They are normally 1ϕ, air-cooled and are manufactured in sizes up to 5 kVA.

Bell-Ringing and Chime Transformers

Bell-ringing and chime transformers are specially designed, small-capacity, constant-potential transformers used to operate doorbells, buzzers, chimes, annunciators, fire or burglar alarms, and industrial signaling. They function with normal primary voltages of 120 V or 240 V and with single or multiple secondary voltages of from 8 V to 24 V.

Neon Sign Transformers

Neon sign transformers operate gas-discharge lamps commonly used with electric signs. Neon sign transformers are varying voltage transformers with windings designed so that the secondary voltage decreases with the load. See Figure 3-24. The center coil is the primary and the two outside coils are the secondary. The secondary voltage decreases as the load increases because of the increased leakage flux and greater voltage drop of the primary. Neon sign transformers are rated according to the secondary voltage and the short-circuit current in milliamperes (mA).

The primary voltage rating is normally 120 V or 240 V while the secondary voltage varies from 2000 V to 15,000 V, depending on the diameter and length of the glass tubing and the type of gas used in the tubing. When the secondary voltage exceeds 7500 V, the two secondary coils normally are connected together and to ground which gives a secondary 1ϕ, three-wire circuit of 7500/15,000 V. The current rating is in the range of from 30 mA to 120 mA. These transformers are self-air-cooled and are normally mounted inside the sheet metal sign enclosure.

NEON SIGN TRANSFORMERS

SECONDARY VOLTAGE =
3000 V TO 15,000 V
AT 30 mA AND 60 mA

LUMINOUS TUBE TRANSFORMER

SECONDARY VOLTAGE =
7500 V TO 15,000 V
AT 120 mA 1800 V
TO 3500 V AT 300 mA

COLD CATHODE TRANSFORMER

SECONDARY VOLTAGE =
5000 V TO 9000 V
AT 30 mA

INDIVIDUAL ILLUMINATED LETTER TRANSFORMER

SECONDARY VOLTAGE =
2000 V TO 7000 V
AT 20 mA

UNENCLOSED TRANSFORMER

Jefferson Electric Co.

Figure 3-24. Neon sign transformers are varying voltage transformers with windings designed so that the secondary voltage decreases with the load.

Special Purpose and Electronic Transformers

Specially designed transformers may be used for a number of purposes other than stepping voltages up or down or for current regulation. Special purpose and electronic transformers are small and are found in all types of electronic, telecommunications, and radio and television equipment. Some of these transformers are used to isolate a piece of equipment from ground or another circuit.

Certain transformers may be used to isolate or to pass certain frequencies such as the RF and IF transformers found in most radios. These are step-down transformers used to match the impedance of two circuits. See Figure 3-25. High-voltage transformers used in TVs are step-up transformers and audio transformers are step-down transformers used to match the output impedance of the output amplifier to speakers.

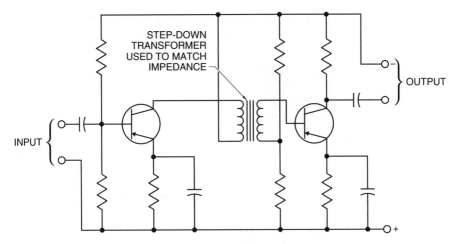

Figure 3-25. Electronic transformers may be used to match the impedance of two circuits.

Two transformers with adjustable metallic core coils may be used in a circuit to make the inductance resonant at a certain frequency. See Figure 3-26. While the design and size of these transformers may be different, they all operate on the same principles as large transformers.

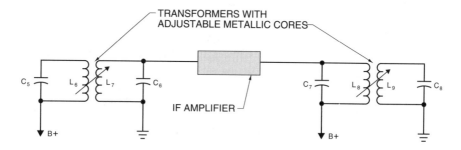

Figure 3-26. Two transformers with adjustable metallic core coils may be used in a circuit to make the inductance resonant at a certain frequency.

Generator Step-Up Transformers

Electrical power is transmitted from a steam, nuclear, or hydroelectric station to the ultimate consumer. See Figure 3-27. The alternators at the steam or hydroelectric generating plants produce power at a nominal voltage in the range of 13.8 kV to 22 kV. Transmission transformers step up this voltage to 115 kV to 750 kV at which the power is transmitted to a central substation where the power from the generating plants is brought together. Large step-down transformers reduce this voltage to a lower voltage in the range of 23 kV to 115 kV as the electrical power nears the city. The feeder lines deliver the electrical power to a distribution substation where the voltage is stepped down to a voltage in the range of 4.2 kV to 25 kV.

The electrical power is delivered to the ultimate consumer by distribution lines at 4.2 kV to 25 kV where it is stepped down by distribution transformers to 115/230 V for residential and small commercial buildings. Some large industrial plants receive electrical power at a voltage of 13.8 kV to 69 kV where in-plant distribution transformers step down the voltage to the desired level for in-plant use.

Generator step-up transformers (power transmission transformers) step up the generator voltage to a high voltage for economical power transmission purposes. Generator step-up transformers for electrical power transmission are installed at the generating plant or source of power. The standard nominal voltage ratings for generator step-up transformers are 13.8 kV, 23 kV, 46 kV, 69 kV, 115 kV, 230 kV, 345 kV, 500 kV, 765 kV, 1000 kV, and 1300 kV. Voltages of 345 kV and higher are classified as extra high voltage (EHV) and are used to transmit power over long distances.

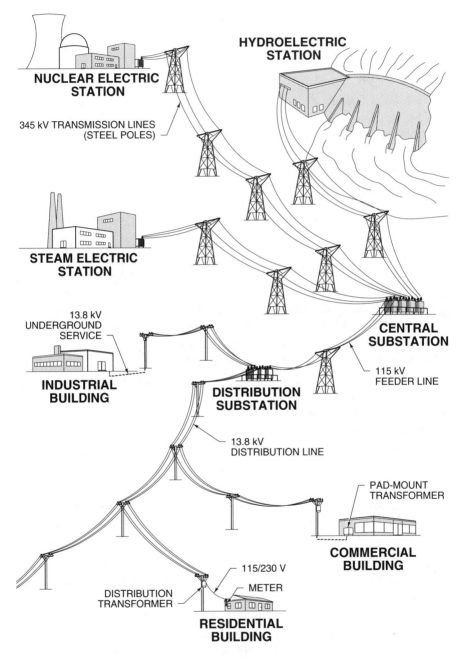

HYDROELECTRIC STATION

NUCLEAR ELECTRIC STATION

345 kV TRANSMISSION LINES (STEEL POLES)

STEAM ELECTRIC STATION

13.8 kV UNDERGROUND SERVICE

INDUSTRIAL BUILDING

DISTRIBUTION SUBSTATION

CENTRAL SUBSTATION

115 kV FEEDER LINE

13.8 kV DISTRIBUTION LINE

PAD-MOUNT TRANSFORMER

COMMERCIAL BUILDING

115/230 V

METER

DISTRIBUTION TRANSFORMER

RESIDENTIAL BUILDING

Figure 3-27. Electrical power is transmitted from a steam, nuclear, or hydroelectric station to the ultimate consumer.

The kVA and kV ratings of generator step-up transformers are largely determined by the amount of power produced by the generator and the distance that the power must be transmitted. Although these units are manufactured in very large sizes, most utilities limit the rating to 500 MVA (500 megavolt-amps = 500,000,000 VA). The reason for this limit is the physical size of one large transformer. Also, if only one large transformer is installed and a malfunction develops, the entire system would be without power. If there are two smaller units and one malfunctions, the other unit could continue to supply power to the system at a reduced value.

Substation Transformers

Substation transformers are normally used to transform the transmission voltage to the voltage level of the primary feeders. There is little electrical difference between a substation transformer and a distribution transformer. A *substation* is an assemblage of equipment installed for switching, changing, or regulating the voltage of electricity. Most manufacturers do not actually define a substation transformer until its intended purpose and the component equipment that is to be used with the transformer has been determined. If a transformer with its associated equipment is to be used by a utility to step down transmission voltages to an intermediate voltage, it is considered a substation transformer.

On the other hand, if a transformer and its associated equipment were installed by a large industrial plant and used to step down a high voltage from a utility transmission line to an intermediate or low voltage, which, in turn, was transmitted and distributed throughout a large area owned and operated by the plant, it would normally be considered an industrial distribution transformer.

A substation is used by a public utility to step down the high voltage of the transmission line to a lower value which is distributed to the consumer. See Figure 3-28. The substation and its associated equipment is manufactured for both indoor and outdoor installations.

General Electric Co.

Figure 3-28. A substation is used by a public utility to step down the high voltage of the transmission line to a lower value which is distributed to the consumer.

Mobile Substations. Mobile substations are used to restore service quickly in emergencies, such as floods, fires, storms, and lightning disasters. See Figure 3-29. Using a mobile substation to bypass a substation also permits planned maintenance during regular hours with no loss in service and with greater safety to personnel. Mobile substations rated up to 25,000 kVA and mobile transformers rated up to 57,000 kVA are used. Mobile substations also permit a higher quality of maintenance.

Distribution Transformers

Distribution transformers change the primary feeder voltage to the voltage required by the consumer. Distribution transformers take power from the primary distribution lines and deliver it directly to the consumer. Because of the increased demand for electrical power and the recent technical advances in the development of special high-voltage cable (in the 25 kV range), the use of high-voltage feeders in industrial plants is becoming more common. Distribution transformers step down the high voltage on the feeder to the value required for the loads. The advantage of this type of electrical power distribution system is that large amounts of power can be delivered to different areas in the plant at a high voltage and a low current, thus resulting in a much smaller size feeder conductor. See Figure 3-30.

Minnkota Power Cooperative, Inc.

Figure 3-29. Mobile substations are used to restore service quickly in emergencies, such as floods, fires, storms, and lightning disasters.

Many public power utilities limit the size of pad-mounted transformers that supply power to consumers to 1000 kVA at 120/208 V and 2000 kVA at 277/480 V. Most public utilities classify transformers operating under 25 kV as distribution transformers. However, transformers could function as distribution transformers at 69 kV for industrial customers provided that the rating does not exceed 500 kVA.

The construction of the outside metal enclosure of distribution transformers depends on the method of mounting and the location of installation. They may be installed indoors or exposed to the weather, installed on a pole, in a special enclosed area either inside or outside a building, in a manhole or vault, on a pad outside the building, on a switchboard, or in other special locations.

Transformers, which are auxiliary to motors and generators, form a subdivision of the distribution class and cover a wide range as to type of design and rating. Those built for starting duty, because the starting cycle covers a relatively short period of time, are physically small compared to their rating. Some of the small low-voltage units are cooled by air, while oil is used for insulating and cooling purposes in units with high voltage and large output ratings.

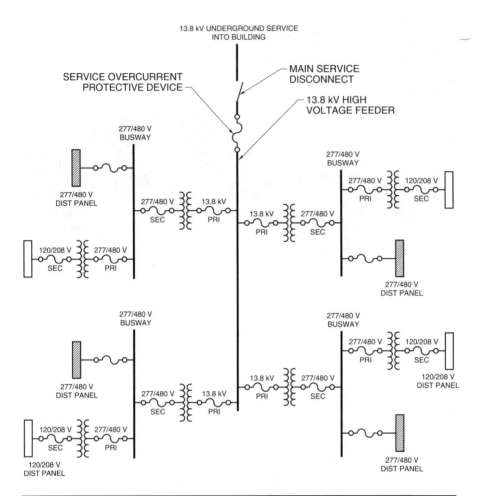

Figure 3-30. Distribution transformers step down the high voltage on the feeder to the value required for the loads.

PURPOSE CLASSIFICATION

Transformers can be classified into four classes according to the current or voltage transformation for which they are employed. The four classes include current, constant-current, constant-potential, and varying-potential transformers.

Current Transformers

Current transformers are designed to change the current of a system. The primary winding is connected in series with a main circuit whose current value requires changing. Common applications of current transformers include street-lighting installations and the operation of instruments.

Constant-Current Transformers

Constant-current transformers are designed to supply a constant current to the secondary circuit. The primary is supplied by a constant voltage source. The voltage of the secondary increases or decreases according to the load, but the secondary current remains constant.

Constant-Potential Transformers

Constant-potential transformers are the most commonly used transformer. Constant-potential transformers are designed to operate at a reasonably constant primary and secondary voltage from no load to full load. Transmission and distribution transformers are examples of constant-potential transformers. Constant-potential transformers are used for the transmission and distribution of electrical power.

Varying-Potential Transformers

Varying-potential transformers are necessary for specific uses, such as for gas-discharge lamps where it is necessary to reduce the secondary with an increase of load. The primary is connected to a constant potential supply circuit normally in the range of 120 V or 240 V.

INSTALLATION REQUIREMENT CLASSIFICATION

Transformers may be classified with respect to the service or purpose for which they are used. Transformers may also be classified with respect to installation requirements. Transformers classified with respect to installation requirements include switchboards, indoor locations, outdoor locations, underground manholes or vaults, residential subsurface transformers, and direct-buried transformers for underground residential systems.

Switchboards

Transformers used with switchboards or panelboards fall into two groups: current transformers and voltage transformers. They are used for the operation of instruments, such as ammeters, voltmeters, wattmeters, and relays for various protective purposes. Such transformers are normally of the dry-type which operate without oil and are normally mounted on the inside of the switchboard enclosure. See Figure 3-31. When higher voltages are involved, these transformers are immersed in oil for insulating purposes.

General Electric Co.

Figure 3-31. Transformers used with switchboards operate ammeters, voltmeters, wattmeters, and relays.

Indoor Locations

Transformers for indoor service are very similar to outdoor transformers. The main difference is that they are not watertight because they are not subject to the elements of nature, such as rain and snow. See Figure 3-32. Requirements are contained in the *National Electrical Code® (NEC®)* (NFPA No. 70) and are mandatory when transformers are installed indoors.

Dry-type transformers may be installed indoors when they are rated at not more than $112\frac{1}{2}$ kVA and not more than 600 V, and the transformer is completely enclosed except for ventilating openings. Dry-type transformers rated at more than $112\frac{1}{2}$ kVA may be installed indoors if they are installed in a transformer room of fire-resistant construction, unless they are constructed with Class B (80°C rise) or Class H (150°C rise) insulation and are properly separated (6′ horizontally and 12′ vertically) from combustible material, or are separated from combustible material by a fire-resistant heat-insulating barrier. All dry-type transformers that are rated at more than 35,000 V are required to be installed in an approved vault. See Figure 3-33.

ABB Power T&D Company Inc.

Figure 3-32. Transformers for indoor service are not watertight because they are not subject to rain and snow.

Askarel, as used in the NEC®, is a synthetic, nonflammable insulating liquid which, when decomposed by the electric arc, involves only nonflammable gaseous mixtures. Askarel is the proper name for non-combustible (fire-resistant) liquid insulation. New askarel is clear and water-white to light straw in color. It has a minimum dielectric strength of 35 kV, a minimum resistivity of 100 Ω/cm × 10^9, and a maximum acid (neutralization) number of 0.014.

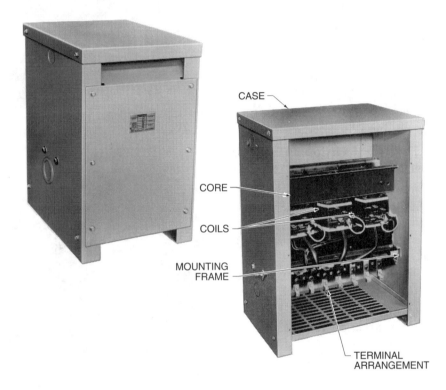

CASE

CORE

COILS

MOUNTING
FRAME

TERMINAL
ARRANGEMENT

ABB Power T&D Company Inc.

Figure 3-33. Dry-type transformers may be installed indoors when they are rated at not more than 112½ kVA and not more than 600 V, and the transformer is completely enclosed except for ventilating openings.

There are other insulating oils used in transformers that also meet the same specifications as askarel that are permitted to be used. Transformers installed indoors containing these types of insulating oil, which are rated in excess of 25 kVA, shall be provided with a pressure-relief vent. When transformers are placed in a poorly ventilated place, special means must be provided for absorbing or removing any gases generated by arcing inside the case. This may be accomplished by providing a means for absorbing any gases generated inside the case of the transformer or by connecting the pressure relief vent to a chimney or flue which carries such gases outside the building. All askarel-insulated transformers that are rated for more than 35,000 V shall be installed in an approved vault.

Special precautions must be taken when oil-insulated transformers are installed indoors because this type of oil is flammable. The NEC® requires that this type of transformer shall be installed in a specially constructed transformer vault that serves no other purpose.

If the total rating of the transformer does not exceed 112½ kVA, the transformer vault may be constructed of not less than 4″ of reinforced concrete, which is less than the requirements for vaults that contain transformers of over 112½ kVA ratings. A vault is not required for oil-insulated transformers installed indoors if they function at a voltage of not more than 600 V and the total rating does not exceed 10 kVA.

Oil-insulated transformers that operate at not more than 600 V and have a total rating not exceeding 75 kVA may be installed inside a building without further fire protection where the surrounding structure of the building is classified as fire-resistant construction. For further information on the installation of transformers indoors, consult Article 450 of the NEC®.

Outdoor Locations

Transformers for installation in outdoor locations are constructed with different types of metal enclosures depending on the requirements of installation. They may be divided into several different groups, such as pole-mounted, pad-mounted, and outdoor protected by a fence or other effective enclosures.

Pole-Mounted. A pole-mounted transformer is used primarily for direct distribution of electrical power to the consumer. When a single transformer is mounted on a pole, the transformer is either secured directly to the pole or supported by a separate mounting bracket which is secured to the pole. The separate mounting bracket is widely used because the transformer can be easily removed for replacement or servicing. Single transformers in the range of 500 kVA can be satisfactorily mounted on a single pole by this method. When several transformers are mounted on a pole, they are supported by a cluster mounting hanger which is secured to the pole. See Figure 3-34.

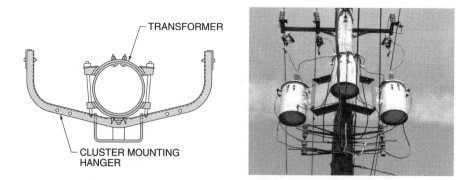

Figure 3-34. Several transformers may be mounted on a pole using a cluster mounting hanger which is secured to the pole.

Three 1ϕ transformers of 250 kVA each can be satisfactorily mounted in this manner. The old method of placing several larger transformers on a platform supported by two or more poles is being largely replaced by cluster-mounted transformers on a single pole or by a pad-mounted transformer for large transformers.

Pad-Mounted. Pad-mounted distribution transformers are becoming extremely popular because of the need for large transformers and the common use of underground services supplied by the utilities. Pad-mounted transformers are designed for mounting directly on a concrete pad outside a building. A pad-mounted transformer is completely enclosed. Pad-mounted transformers are primarily used for the direct distribution of electrical power from the utility to the consumer. These transformers are available in small sizes in the 10 kVA range for residential use and in sizes in the 5000 kVA range for commercial and industrial use. See Figure 3-35. Pad-mounted transformers are available over a wide range of kVA ratings and are either air- or oil-cooled, depending on the size and voltage.

Outdoor Protected By Fence. Outdoor installation of transformers together with the necessary switching equipment and protective devices which comprise an outdoor substation are commonly used for distribution and power transmission. When transformers and their associated equipment are installed outdoors, the installation must be weatherproof and, if not totally enclosed, must be inaccessible to unauthorized personnel.

PAD-MOUNTED TRANSFORMERS

General Electric Co.

Figure 3-35. Pad-mounted transformers are designed for mounting directly on a concrete pad outside a building.

For small-capacity installations, the transformer and its associated equipment may be contained in a common weatherproof enclosure with no live parts exposed. For large installations where some hazards would exist if the installation were accessible to the public, a barrier or fence must be installed to eliminate the entrance of all but authorized personnel.

Transformers that are installed outdoors are oil-insulated. Special precautions must be taken because the oil is flammable. The NEC® requires that oil-insulated transformers shall be installed and located so that a fire originating in the transformer will not spread to adjacent buildings or combustible material or will not, in any way, block or conflict with doors, windows, or fire escapes. Special safeguards that are recognized by the code are: space separation, fire-resistant barriers, automatic water spray systems, and special enclosures which confine the oil of a ruptured transformer tank. Oil enclosures may consist of curbed areas or basins, fire-resistant dikes, or trenches filled with coarse crushed stone.

Underground Manholes or Vaults

Distribution systems are installed underground in areas where space is limited. Subsurface transformers are specially designed for underground installation to serve underground distribution systems. Such transformers are subject to complete submergence in water and are constructed so that all joints (such as between the tank and cover bushing, and the tank wall or bushing and cover) are watertight under varying heads of water pressure.

A submersible distribution transformer is used for underground residential distribution systems. See Figure 3-36. Installed in a below-grade vault, it is completely safe

General Electric Co.

Figure 3-36. A submersible distribution transformer is used for underground residential distribution systems.

and out of sight. A submersible distribution transformer, normally referred to as a subway transformer, is used for underground commercial distribution systems. In large cities, where large amounts of power are required in small areas, transformers may be mounted in underground vaults in the same area where the power is used. See Figure 3-37.

General Electric Co.

Figure 3-37. Transformers may be mounted in underground vaults in the same area where the power is used.

To maintain constant service, a number of network transformers in various parts of the city may be connected in a parallel network. Each unit is provided with a short-circuit switch which short circuits and grounds the primary winding while the transformer is disconnected from the circuit. The switch enables the circuit to be isolated from the other circuits in the network if a fault occurs in the other circuits.

A network system consists of a grid of interconnecting cables operating at utilization voltage. The grid is energized at many points so that the loss of any one point of supply does not cause loss of service. When a source is lost, the load formerly supplied at that point is absorbed by the other sources of power. See Figure 3-38.

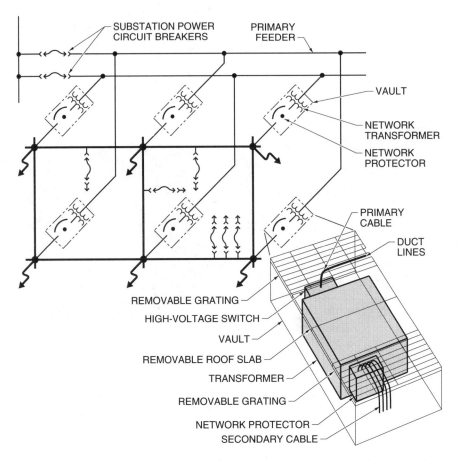

Figure 3-38. A network system consists of a grid of interconnecting cables energized at many points so that the loss of any one point of supply does not cause loss of service.

Residential Subsurface Transformers (RSTs). A residential subsurface transformer (RST) is designed to provide economical, reliable distribution service primarily for residential areas. Installed in an underground vault, it is completely out of sight. The Leapfrog® II RST is designed with the necessary cooling surface to operate at 65°C in an enclosure when used with an efficient air-directing baffle and operating at rated load. This transformer is available for 1φ power in ratings of 15 kVA through 167 kVA at a normal voltage of 15 kV and below. See Figure 3-39.

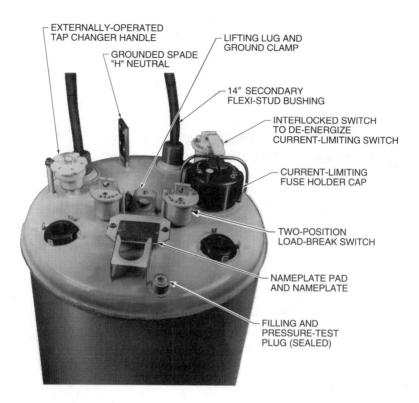

EXTERNALLY-OPERATED
TAP CHANGER HANDLE

GROUNDED SPADE
"H" NEUTRAL

LIFTING LUG AND
GROUND CLAMP

14″ SECONDARY
FLEXI-STUD BUSHING

INTERLOCKED SWITCH
TO DE-ENERGIZE
CURRENT-LIMITING SWITCH

CURRENT-LIMITING
FUSE HOLDER CAP

TWO-POSITION
LOAD-BREAK SWITCH

NAMEPLATE PAD
AND NAMEPLATE

FILLING AND
PRESSURE-TEST
PLUG (SEALED)

General Electric Co.

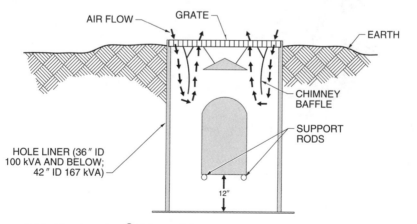

AIR FLOW

GRATE

EARTH

CHIMNEY
BAFFLE

SUPPORT
RODS

HOLE LINER (36″ ID
100 kVA AND BELOW;
42″ ID 167 kVA)

12″

Figure 3-39. The Leapfrog® II is an RST designed to provide economical, reliable distribution service for residential areas.

The specially designed chimney baffle uses the natural chimney effect that causes cool outside air to enter on the outside of the vault where its greater density carries it to the bottom of the vault. The air flows around the transformer's surface, picking up heat as it rises through the inner opening of the baffle.

Direct-Buried Transformers for URD (Underground Residential Distribution) Systems. Direct-buried transformers for URD systems are constructed so that all joints are watertight and may have a polymer or mild steel tank. See Figure 3-40. In the polymer-tank direct-buried transformer, the tank and cover are molded from reinforced polyester with the high-voltage bushing well, high-voltage neutral, and low-voltage bushings molded into the cover. The polymer tank is impervious to corrosion attacks from scrapes, scratches, or bumps that may occur during installation.

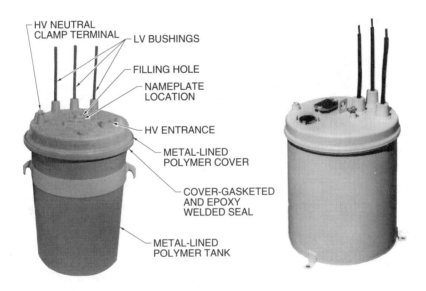

General Electric Co.

Figure 3-40. Direct-buried transformers for URD systems are constructed so that all joints are watertight and may have a polymer or mild steel tank.

In the mild steel tank direct-buried transformer, the mild steel tank has a superior paint finish that is grit-blasted, passivated, epoxy-primed, and painted with the finish coat. The unit is then baked. The epoxy finish has superior abrasive resistance and excellent chemical and temperature stability. Both the polymer tank and the mild steel tank units are hermetically sealed, oil-filled units. For further information on the installation of transformers outdoors, consult the NEC®.

PHASE CLASSIFICATION

All transformers, without regard to service applications, are divided into 1ϕ and 3ϕ. In 1ϕ transformers, although there may be several magnetic paths in parallel (for example the distributed shell-type magnetic circuit), all the fluxes in these paths are induced by the same voltage and are therefore in phase with each other. There may be two or more windings on the magnetic circuit, but the voltages induced and the currents flowing are 1ϕ currents because as they are all produced by the 1ϕ flux.

In a 3ϕ transformer, there are three flux paths. Each of the fluxes in these paths is displaced from the others by 120°, which is the 3ϕ relationship of the voltages impressed upon the primary windings. Parts of these magnetic paths may be common. Any two of the 3ϕ fluxes may use the same path, but they must function magnetically as separate magnetic paths.

A 3ϕ transformer has three similar primary and secondary windings, one primary and one secondary winding for each phase of the 3ϕ primary supply voltage and secondary voltage. A 3ϕ transformer is essentially three 1ϕ transformers built on one magnetic circuit. In some cases, a 3ϕ transformer may actually be an assembly of three separate 1ϕ transformers in one tank, with the primary and secondary coil leads of each unit arranged on a terminal board so that a 3ϕ connection is obtained with only three leads from the primary and secondary sides of the terminal board.

The majority of transformers made in the U.S. are 1ϕ. One reason for the large number of 1ϕ transformers is their use for residential consumers. Another contributing factor is that when 3ϕ power is supplied to a consumer from an overhead distribution line, three 1ϕ distribution transformers would normally be mounted on a pole instead of one large 3ϕ transformer because of the ease of handling and mounting the three 1ϕ units. Another advantage of using three 1ϕ transformers to feed a 3ϕ lighting and motor load is where the 1ϕ lighting load is much larger than the 3ϕ motor load (such as an office building where the only 3ϕ load is air-conditioning). In this type of installation, one large transformer would be installed to serve the lighting and air-conditioning load and the other two transformers need only be large enough to serve the 3φ load for the air-conditioning.

A 3φ transformer is normally used when transformers are used to transform 3φ power (with the exception of pole-mounted transformers). A 3φ transformer requires three times as much copper as a 1φ unit but, in comparison with three 1φ units, a 3φ transformer requires less than three times as much iron for the core.

In comparing three 1φ units to a 3φ transformer, the 3φ transformer is lower in initial cost, costs less to install, has simpler connections, a higher efficiency, and requires a much smaller space. See Figure 3-41.

TRANSFORMER ADVANTAGES	
3φ	1φ
Low initial cost	Inexpensive spare parts
High efficiency	Low repair cost
Less total weight than 1φ	More voltage than 3φ
Less total floor space than 1φ	Greater flexibility—a single unit can be replaced in case of trouble
Low installation cost	
Low transportation cost	

Figure 3-41. A 3φ transformer is lower in initial cost, costs less to install, has simpler connections, a higher efficiency, and requires a much smaller space than three 1φ transformers.

Transformer Standards

If cost were not a factor, all transformers would be designed for the specific power system and load that the transformer serves. It is uneconomical to design all transformers specifically for each power system and load served. It has become a universal practice to standardize many transformer designs.

Many construction details are standardized to make it easier for the users of transformers. Ratings have been standardized in great detail, especially for transformers in the small power and distribution class. The guiding principle of standardization is to make the standard specification such that the total cost of manufacturing and using the transformers is relatively low.

Standard Kilovolt-Amp Ratings

Standard ratings have been developed for distribution and power transformers. In most cases, the ratings of 3ϕ transformers equal the ratings of three 1ϕ transformers grouped in a 3ϕ bank. See Figure 3-42. While these are preferred ratings, 1ϕ distribution transformers may be found as low as 10 kVA and 3ϕ distribution transformers as low as 30 kVA.

PREFERRED 1ϕ AND 3ϕ DISTRIBUTION AND POWER TRANSFORMER kVA RATINGS			
1ϕ	1ϕ Continued	3ϕ	3ϕ Continued
25	4000	75	10,000
37.5	5000	112.5	12,000
50	6667	150	15,000
75	8333	225	20,000
100	10,000	300	25,000
167	12,500	500	30,000
250	16,667	750	37,000
333	20,000	1000	50,000
500	25,000	1500	60,000
667	33,000	2000	75,000
833	40,000	2500	100,000
1000	50,000	3000	120,000
1250	66,670	3750	150,000
1667	83,333	5000	200,000
2000	100,000	6000	250,000
3333	—	7500	300,000

Figure 3-42. In most cases, the ratings of 3ϕ transformers equal the ratings of three 1ϕ transformers grouped in a 3ϕ bank.

Basic Impulse Insulation Level (BIL)

The insulation of a transformer is subject from time to time to momentary overvoltages that may be caused by system faults, switching surges, or lightning strikes. The calculation of probable overvoltages or impulse voltages actually appearing at the transformer is determined from the characteristics of the system with the elements of nature, such as lightning, taken into consideration. Normally, overvoltages caused by lightning are of sufficient magnitude to flash over or break down transformer insulation and therefore create the greatest potential hazard to transformer insulation.

At or near a lightning strike, the voltage increases very rapidly until it reaches its peak, perhaps to 1,000,000 V in $1/1,000,000$ sec (1000 kV per second). Special precautions are normally taken which limit the lightning voltage which could appear at the transformer terminals. These precautions normally consist of overhead ground wires which shield the overhead line conductors, reduced line insulation or gaps near the transformer so that the lightning flashes over to limit the voltage, and lightning arresters installed as near as possible to the transformer which discharge the lightning impulses to ground. Transformer insulation must be able to withstand the normal operating voltage, the maximum fault voltage, and the probable impulse voltage.

The total voltage that could appear at the terminals of a transformer is determined from the characteristics of the system. It is impractical to design each transformer to withstand the voltages which may reach it, especially as the impulse voltages, which can reach the transformer terminals, are different for each location on the system.

All transformers are designed to withstand one of a number of standard impulse tests. From consideration of the voltages, which may actually appear at the transformer, one of the standard test strengths (basic insulation level) is chosen. The basic insulation level for the transformer is selected so that its strength is 15% to 20% above the impulse voltages which can reach it.

Forms of Construction

The two general forms of construction are the shell form and the core form. While there are several modifications of these two forms, they are all classified as either one. In the shell form, the magnetic circuit surrounds the current circuit. In the core form, the current circuit surrounds the magnetic circuit. See Figure 3-43.

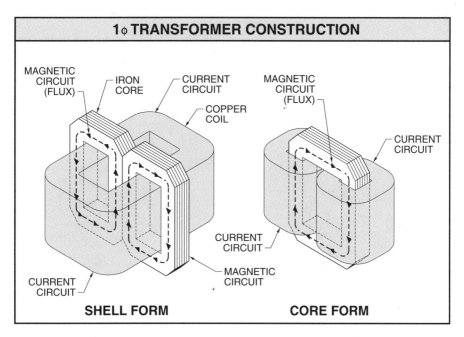

Figure 3-43. The two general forms of construction are the shell form and the core form.

The shell and core forms of construction each have their own inherent characteristics which make them especially adaptable to certain voltage and capacity ratings. When considering which form to use, the factors considered include:

- Economy of material for a given performance

- Adaptability to heavy currents and high voltages

- Ventilation or cooling

- Mechanical stresses

- Repairs

For each voltage and frequency class there is a transformer of a certain output rating, above which the space factor makes the shell form cheaper, and below which the core form uses less material. In general, small and medium power transformers are of core form construction and large power transformers are of the shell form. Three-phase transformers may have a rectangular core magnetic circuit or a cross-shaped core magnetic circuit. See Figure 3-44.

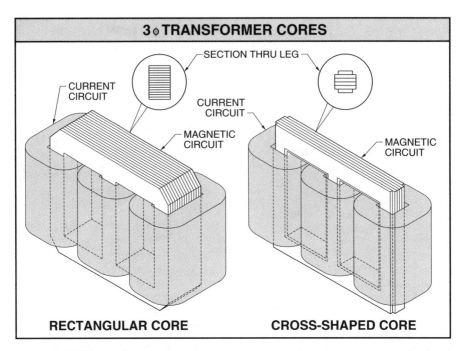

Figure 3-44. Three-phase transformers may have a rectangular core magnetic circuit or a cross-shaped core magnetic circuit.

Tertiary Winding

A tertiary winding is an auxiliary delta-connected winding, particularly in wye-wye connected transformers. This stabilizes the neutral point of the fundamental frequency voltages, protects the transformer and the system from excessive third harmonic generated voltages, and prevents telephone interference due to third harmonic currents and voltages in the lines and earth.

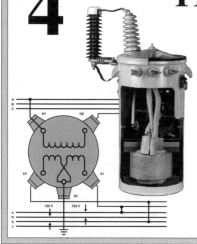

4 TRANSFORMER CIRCUITRY

Transformer circuits consist of magnetic and current circuits. The magnetic circuit is the path of flux lines produced by the current in the primary winding. The current circuit comprises the primary and secondary windings.

MAGNETIC CIRCUITS

The magnetic circuit in a transformer provides a path of low resistance to the flow of magnetic flux lines produced by the current in the primary or exciting winding. Magnetic circuits may be composed of a coil with an air core, a coil with an iron core, and a coil with an iron core and air gap. See Figure 4-1.

If a voltage (E) is impressed across each coil, a current is set up in each coil in the direction of the arrows on the leads. This current causes a flux to flow in the magnetic circuit (dashed lines). To calculate magnetic flux in a circuit, apply the formula:

$$\Phi = \frac{.4\,\pi\,NI}{l/_{\mu\,A}}$$

where

Φ = magnetic flux

N = number of turns in coil

I = current through coil

l = mean length of magnetic path

μ = permeability of material in magnetic circuit

A = cross-sectional area of magnetic path

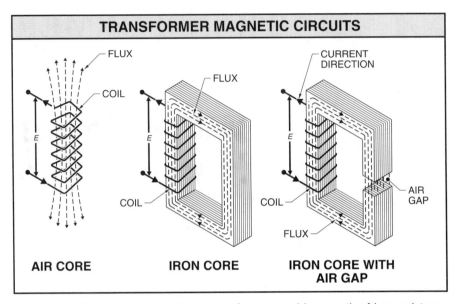

Figure 4-1. The magnetic circuit in a transformer provides a path of low resistance to the flow of magnetic flux lines produced by the current in the primary winding.

Example: Calculating Magnetic Flux

What is the magnetic flux of an iron and air core magnetic circuit that has 100 A flowing through 1000 turns? The cross-sectional area of the core is 1 cm² (1 cm² = 10^{-4} m²), the iron core has a permeability of 4000, the average path length of the circuit is 24 cm, and there is a 1 cm air gap in the core. *Note:* Total reluctance ($^1/_{\mu A}$) in ampere-turns per Weber (At/Wb) is determined by calculating the reluctances for the iron (R_I) and air gap (R_G).

1. Calculate reluctance of air gap.

$$R_G = \frac{1}{\mu A}$$

$$R_G = \frac{0.01 \text{ m}}{4\pi \times 10^{-7} \times 10^{-4} \text{ m}^2}$$

$$R_G = 7.95 \times 10^7 \text{ At/Wb}$$

2. Calculate reluctance of iron.

$$R_I = \frac{1}{\mu A}$$

$$R_I = \frac{0.23\ m}{1.6\pi \times 10^{-3} \times 10^{-4}\ m^2}$$

$$R_I = 4.56 \times 10^5\ At/Wb$$

3. Calculate total reluctance.

$$R_T = 795 \times 10^5 + 4.56 \times 10^5$$

$$R_T = 799.6 \times 10^5\ At/Wb$$

4. Calculate magnetic flux.

$$\Phi = \frac{.4\pi NI}{1/\mu A}$$

$$\Phi = \frac{.4 \times 3.14 \times 100 \times 1000}{799.6 \times 10^5}$$

$$\Phi = \frac{125,600}{79,960,000}$$

Φ = 0.00157 Webers

As the ampere-turns (*NI*) of the exciting winding are varied, either by varying the turns, the current, or both, the flux varies in each of the three magnetic circuits. The permeability (μ) of air is constant and the flux is directly proportional to the ampere turns of magnetizing force. The number of flux lines per unit of cross-sectional area of the magnetic circuit (flux density) is proportional to the magnetizing force. Magnetizing curves show the value of flux density for varying values of ampere turns of magnetizing force. See Figure 4-2.

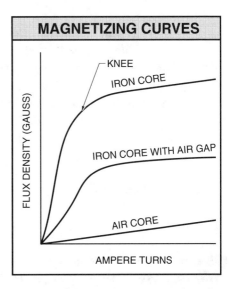

Figure 4-2. Magnetizing curves show the value of flux density for varying values of ampere-turns of magnetizing force.

The magnetizing curve for an air core is a straight line. The permeability of iron is much greater than the permeability of air. Therefore, for a given number of ampere turns, the flux density in the iron circuit is much greater than the flux density of air. The number of ampere turns required to produce the total flux is much smaller in the iron circuit than in the air circuit for a given flux density. The permeability of iron is not constant and as the flux density in the iron circuit is increased, the iron becomes saturated and the curve for the iron circuit flattens, becoming parallel to that of the air core. At very high flux densities, the iron core magnetic circuit is only slightly better than the air core magnetic circuit.

If an air gap is introduced in the iron core, the reluctance of the magnetic path increases greatly. The magnetization curve becomes straight and lies between the magnetization curves for iron and air. The greater the number of air gaps, the straighter the curve becomes, moving closer to the curve of an air core. Reactance coils for transmission lines are made with an air core because the straight line characteristic is especially desirable at high flux densities.

Construction

In a transformer, the magnetic circuit is made entirely of iron. The iron, in the form of laminated sheet steel strips, carries the required flux with a minimum amount of magnetizing force or ampere turns. The steel strips are cut from silicon-iron sheets most commonly having a thickness of 14 mils. Both sides of the steel strips are coated with an enamel, forming an insulating coating (lamination) approximately $\frac{1}{2}$ mil thick. The length of the lamination is in the direction of magnetic flow.

Aging and Annealing

The sheet steel used in today's transformers contain a sufficient amount of silicon to prevent the steel from aging. Aging is the characteristic of the magnetic circuit which increases the iron loss as the transformer continues to operate over a period of time. Before the introduction of silicon in the steel, all steel showed signs of aging. Sheet steel, as it comes from steel mills, has a very high loss and low permeability. These two factors are greatly improved by the annealing process.

During the punching or shearing of the sheet steel into laminations of the desired length and width, strains are set up to increase the loss and decrease the permeability. If low losses are desired, it is necessary to anneal the sheet steel after the punching or shearing operation to relieve these strains. Hammering or bending the lamination to give it a permanent set also sets up stresses which increase the iron loss. Therefore, when a transformer is dismantled for repairs, it is often necessary to anneal the laminations before rebuilding the transformer if the iron loss after rebuilding is not to exceed the original loss of the transformer.

Exciting Current

Exciting current is the current that the magnetic circuit draws from the line when it is excited at normal voltage and frequency with no load on the secondary. The current may be measured by an ammeter placed in the circuit.

The exciting current represents a flow of current into and out of the magnetic circuit. This current is objectionable because it produces a loss in the generator supplying the current and a loss in the lines leading to the transformer.

The total ampere turns of magnetizing force are equal to the exciting current (in amperes) multiplied by the number of turns in the excited winding. Therefore, for a given transformer having a fixed number of turns, the total flux can be varied only by a change in exciting current. The exciting current at high flux densities increases rapidly because the iron circuit becomes saturated. For this reason, the magnetic circuit must be worked at a density low enough to limit the exciting current to a reasonable value and at the same time permit the transformer to operate at an increased voltage without greatly increasing the current.

Iron Loss

While measuring the exciting current of a transformer, the iron loss is the power (in watts) supplied to the transformer as measured by a wattmeter connected in the iron circuit. The iron loss is less than the volt-ampere input when compared with the volt-ampere input of the transformer when it is excited at normal voltage with no secondary load.

This indicates that the exciting current is not in phase with the impressed voltage. The ratio of the iron loss (in watts) to the volt-ampere input is the power factor of the transformer at no load. The ratio of the iron loss to the volt-ampere input is a very low value. A transformer with high-exciting current contributes to a poor power factor in the circuit. The iron loss is made up of hysteresis and eddy-current losses.

Hysteresis Loss. The hysteresis loss represents the energy required to reverse the direction of the flux in the magnetic circuit. It is proportional to the weight of iron in the circuit, the frequency of the flux, and approximately the 1.6 power of the magnetic density.

Eddy-Current Loss. The flux changing in the iron circuit generates voltages in the laminated paths which set up a flow of current (eddy current) at right angles to the direction of the flux path. The eddy-current loss is proportional to the total weight of the iron, the square of the frequency of the flux, and the square of the magnetic density.

Variation of Iron Loss and Exciting Current

With other factors constant, the iron loss of a transformer varies with the frequency. The iron loss does not vary directly with the frequency, but at a mathematical power somewhere between the first and second (square of a number) power. This mathematical power depends on the ratio of hysteresis loss to the eddy-current loss. The hysteresis loss changes directly with the frequency, while the eddy-current loss changes directly with the square of the frequency.

For example, the sum of the ratio of the hysteresis loss to the eddy-current loss in a transformer, which equals the total iron loss, varies as the 1.2 power of the frequency. With the frequency constant, the total iron loss varies at a mathematical power of the magnetic density between 1.6 and 2.0. The iron loss varies almost directly with the square of the magnetic density for the magnetic flux densities at which transformers ordinarily operate.

The iron loss varies with a change of frequency if the impressed voltage is constant. For example, if the frequency is reduced from 60 Hz to 25 Hz, the total iron loss is less than half its former value. This is due to the change in frequency if the magnetic density remains constant. The magnetic density increases in the ratio of 60:25 due to the reduced frequency. The total iron loss increases approximately as the square of the magnetic density. The net change is that the iron loss increases and nearly doubles while changing from 60 Hz to 25 Hz.

With the frequency constant, and considering the change in iron loss which is produced by an increase or decrease in the impressed voltage, the iron loss is roughly proportional to the square of the voltage. For example, an increase in the impressed voltage of 10% increases the iron loss approximately 20%. A corresponding reduction occurs when the impressed voltage is lowered.

It is difficult to get a definite idea of the change in the exciting current with variations in frequency and voltage. This is due to the shape of the magnetizing curve. Near the knee of the curve, the change in exciting current is very rapid with a variation of either the frequency or the voltage. An estimate of the change in exciting current with the impressed voltage may be obtained because an increase of approximately 10% of the impressed voltage nearly doubles the exciting current.

CURRENT CIRCUITS

The current circuit of a transformer comprises the primary and secondary windings which may assume various forms. The copper or aluminum conductors used in the primary and secondary windings may have a round, square, or rectangular cross section. Small conductors, up to and including size No. 11 (approximately .0907″ in diameter), are normally round. Square wire is not manufactured in sizes much smaller than .0907″ on a side. The small-sized square wires are difficult to wind because they turn on edge and cut through the insulation between turns and layers. Rectangular conductors of small area are also undesirable because their thin edges often cut through the insulation. Thin conductors are also uneconomical because the insulation on the conductors occupies a greater amount of space than required if the conductor were made thicker and narrower. The maximum size of a rectangular conductor is limited by the ease with which it can be bent or formed into the desired coil shape.

Eddy currents flowing in the conductor create eddy-current losses which cause additional heating and decreased efficiency. These losses are proportional to the square of the thickness of the conductor. The thickness of the conductor depends on the maximum allowable eddy currents in the conductor.

The various types of insulating material used on the winding conductors are determined by the average temperature rise of the aluminum or copper conductors inside the coil windings. The most commonly used coil winding insulations are high-temperature, heat-resistant enamels or specially treated, stabilized paper tapes which are chemically modified to resist aging. Depending on the requirements of a particular design, some winding conductors may require several layers of the stabilized paper tape.

Windings

Assembled transformer windings may be divided into concentric and interleaved windings. Concentric windings have the various coils forming the completed winding assembled concentrically. Interleaved windings are formed by assembling the individual coils side by side. See Figure 4-3.

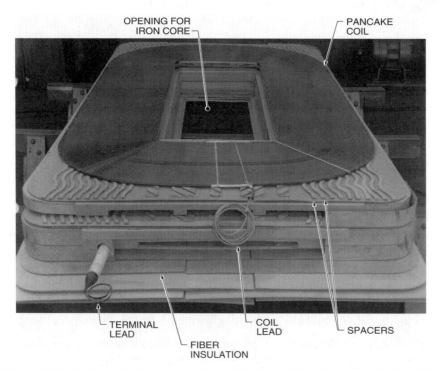

Figure 4-3. Interleaved windings are formed by assembling the coils side by side.

The individual coils forming either the concentric or the interleaved windings may be rectangular or circular in shape. In general, concentric windings are used with core-type construction and interleaved windings are used with shell-type construction. At times, interleaved windings are used for large core designs and concentric windings are used with small shell-type units.

Concentric. In the concentric coil windings, the coil conductors are wound on a tubular form. Two types of concentric coil construction are the cylindrical-wound coil and the continuous-wound pancake coil. A cylindrical-wound coil consists of one or more layers of insulated conductors wound on a tubular form. See Figure 4-4. The windings are wrapped with epoxy glass tape which, after baking, forms a strong band around the coil. This type of coil normally functions at voltages up to 8.7 kV (95 BIL) and currents up to 2500 A.

The conductors of a continuous-wound pancake coil consist of one to five rectangular straps in parallel that are wound continuously on a tubular form. See Figure 4-5. The coils are pressed hydraulically during construction to the calculated short-circuit forces that they must withstand when in service. This type of coil is suitable for substation use and normally functions at voltages above 8.7 kV (95 BIL).

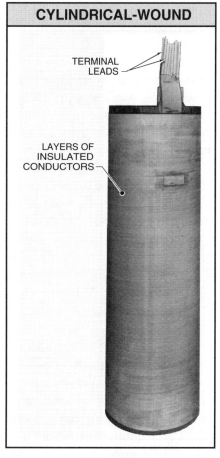

CYLINDRICAL-WOUND

TERMINAL LEADS

LAYERS OF INSULATED CONDUCTORS

ABB Power T&D Company Inc.

Figure 4-4. A cylindrical-wound coil consists of one or more layers of insulated conductors wound on a tubular form.

Interleaved. Interleaved coil windings may be used with either the core- or shell-type of construction. Nearly all large shell-type transformers use interleaved windings. Few core-type transformers use interleaved windings. The individual coils of the interleaved windings are sometimes referred to as pancake coils. Pancake coils are usually wound with square or rectangular conductors and one or more conductors in parallel, depending on the current to be carried. Winding the coils in this form with insulation between the coils exposes the maximum amount of conductor surface to the cooling medium. This enables the heat to be conducted rapidly from the coils to the cooling medium, which prevents overheating.

Individual concentric coils are wound on micarta or heavy pressboard cylinders. The cylindrical-wound coil is normally wound directly on a heavy pressboard cylinder with oil impregnated pressboard insulation between adjacent layers. The continuous-wound pancake coil is normally wound over vertical insulation strips on a heavy pressboard cylinder. The individual

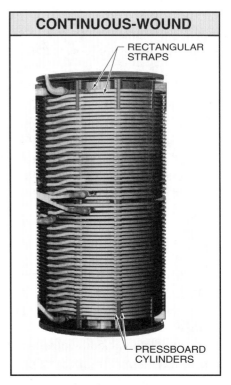

CONTINUOUS-WOUND

RECTANGULAR STRAPS

PRESSBOARD CYLINDERS

ABB Power T&D Company Inc.

Figure 4-5. The conductors of a continuous-wound pancake coil consist of one to five rectangular straps in parallel that are wound on a tubular form.

coils are wound so that when assembled together, there is sufficient space between the coils to form adequate ducts for the circulation of oil for cooling the coil. The core and coil assembly must be rigidly braced to withstand the mechanical forces under line fault conditions and to resist vibration and shock forces during shipment.

Individual pancake coils are assembled with flat insulating barriers between coils, normally with spacers of pressboard to permit the circulation of oil. The individual coils are assembled into high-voltage and low-voltage groups and these groups are assembled together to form the completed high-voltage and low-voltage winding.

Forces On Windings

Moderate axial and radial forces exist between windings in a transformer under normal operating conditions. This force becomes severe when a system short-circuit occurs or when all or part of a winding becomes shorted.

A transformer must be able to withstand faults on a system that is external to the transformer. Standards specify that a fault directly on the bus adjacent to the transformer shall not result in any damage to the transformer. The transformer must be able to carry 10 times the full-load current for periods up to several seconds without damage.

When a transformer's primary is being supplied with power but there is no load on the secondary, the leakage flux is negligible. Nearly all the flux is conducting flux in the iron core linking the two windings. However, as load current flows, leakage flux is present. This leakage flux increases with the load. Under fault conditions, this leakage flux increases to the point where it tries to force the secondary winding into the core. In this case, the primary winding is considered as a circle of wire and the force from the fault is trying to enlarge the circle. In other words, the force is attempting to stretch the copper. Because copper has tensile strength, it resists this force, damaging only the secondary winding. Not only does the force crush the secondary winding, it also drives it downward and drives the primary winding upward. Vertical forces are established if the magnetic centers of the two windings are not in line. The greater the distance that these magnetic centers are offset, the greater the vertical force. Therefore, the magnetic centers should be as nearly in line as possible.

To prevent the unbalance of magnetic centers, it is normal to have taps on one or both windings. These winding taps may be set to remove a portion of the winding so that the magnetic centers of both windings are as close to being symmetrically placed as possible.

Copper Loss

The *copper loss* of a transformer is the energy loss in the windings when the transformer is loaded. Copper loss is composed of two parts. The first is the product of the square of the current flowing in the windings and their resistances. The second is due to the small eddy currents flowing in the copper. The eddy-current loss is due to the magnetic leakage of the flux through the air.

The eddy-current loss may be as high as 15% of the total copper loss. The copper loss due to the resistance of the windings depends on the temperature of the windings because the resistance of the copper varies with temperature. As the temperature rises, the resistance increases. As the temperature falls, the resistance decreases. The eddy-current loss decreases with an increase of temperature because the actual value of the eddy current decreases as the resistance increases. The temperature must be stated at which the copper loss is specified. The total of these two losses may be found by short-circuiting the secondary winding and measuring the input when enough voltage is applied to either winding to cause normal current to flow. The effective resistance value of the transformer is obtained if the input is divided by the square of the primary current flowing.

Impedance voltage is the voltage impressed on the transformer to circulate normal full-load current. The impedance voltage is the sum of the resistance drops through the primary and the secondary windings and the reactive drop. The regulation of the transformer depends on the impedance voltage which is normally expressed as a percentage of the rated primary voltage.

Reactive Voltage

Leakage flux lines which pass only through the primary coil are not effective in inducing voltage in the secondary coil. Likewise, flux lines which pass through the secondary coil and not through the primary are not effective in producing a counter electromotive force opposing the impressed electromotive force in the primary winding. These two elements form the total reactive drop through the transformer winding. The reactive drop may be calculated after the impedance and the copper loss are known.

For example, the secondary of a 100 kVA, 2400 V to 240 V, 1ϕ, 60 Hz transformer is short-circuited. With an ammeter in the primary side, a voltage of 120 V across the primary winding is necessary to send full-load current through the winding. A wattmeter in the circuit reads 3000 W. The copper loss is therefore 3000 W which, when expressed as a percentage of kVA rating of the transformer is 3%. The impedance equals 120 V and, expressed in percentage of normal voltage of winding across which the 120 V is measured, is 5% ($120 \times 100 \div 2400 = 5\%$) impedance. The reactive drop expressed in percent is 4% ($\sqrt{5^2 - 3^2} = \sqrt{16} = 4\%$).

This reactive drop in percent may be expressed in volts by multiplying by the rated voltage. In terms of the primary winding, the reactive voltage drop is 4% of 2400 V or 96 V. In terms of the secondary winding, the reactive voltage drop is 4% of 240 V or 9.6 V. Likewise, the impedance voltage in terms of the secondary winding is 5% of 240 V or 12.0 V.

Impedance

A step-up transformer, while increasing voltage, increases impedance with a resulting decrease in current. Likewise, a step-down transformer, while decreasing voltage, decreases impedance and increases current. Therefore, a transformer changes impedance but the impedance is more pronounced than the voltage change. In fact, a transformer changes impedance by the square of the turns ratio. To calculate impedance, apply the formula:

$$Z_1 \div Z_2 = N_1^2 \div N_2^2$$

where

Z_1 = impedance of primary winding (in Ω)

Z_2 = impedance of secondary winding (in Ω)

N_1^2 = square of the turns in primary winding

N_2^2 = square of the turns in secondary winding

This formula may be rearranged to determine any value when the others are known.

Example: Calculating Impedance

What is the impedance of the secondary winding of a transformer with a turns-ratio of 10:1 when the primary impedance is 50 Ω?

$$Z_2 = \frac{Z_1 \times N_2^2}{N_1^2}$$

$$Z_2 = \frac{50 \times (1)^2}{10^2}$$

$$Z_2 = \frac{50 \times 1}{100}$$

$$Z_2 = .5 \ \Omega$$

Regulation

The regulation of a transformer is the difference in voltage between no load and full load expressed as a percentage of the full-load voltage. This regulation depends on the reactive voltage and the total copper loss of the transformer. When supplying a noninductive load, the copper loss forms the major portion of the voltage drop. As the power factor of the load decreases, the reactive voltage increasingly adds to the total voltage drop. When good regulation is required, it is necessary that the copper loss and reactive voltage be kept to a low value.

Efficiency

The *efficiency* of a transformer is the ratio of the output to the input. It is also equal to the ratio of the output to the output plus losses. That is, efficiency = output ÷ input = output ÷ output + copper loss + core loss.

In a standard power transformer, the efficiency is generally from 96% to 99%. The losses are due to the copper losses in both windings and the hysteresis and eddy-current losses in the iron core. The copper losses vary with the square of the current in the winding and the winding resistance. The core losses, consisting of the hysteresis and eddy-current losses caused by the alternating magnetic flux in the core, are approximately constant from no load to full load with rated voltage applied to the primary.

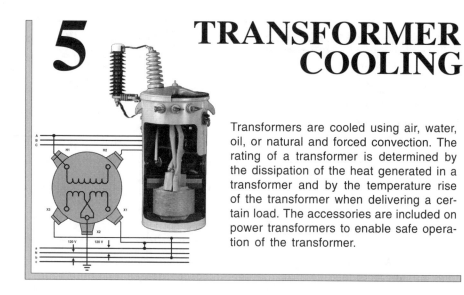

5 TRANSFORMER COOLING

Transformers are cooled using air, water, oil, or natural and forced convection. The rating of a transformer is determined by the dissipation of the heat generated in a transformer and by the temperature rise of the transformer when delivering a certain load. The accessories are included on power transformers to enable safe operation of the transformer.

TRANSFORMER COOLING

The rating of a transformer is determined by the dissipation of the heat losses generated in a transformer and by the temperature rise of the transformer when delivering a certain load. Because of the close concentration of the conductors in the windings, heat is a factor in transformer installation and design. If excessive heat is permitted to remain in or around the transformer, the insulation on the conductors that comprise the windings would be subject to damage and perhaps failure. Any means that can economically dissipate the heat generated in the coils and magnetic circuit more readily permits the increase in the rating of a transformer having a definite physical size or permits a decrease in the dimensions of a unit having a definite kilovolt-ampere rating.

Cooling Abbreviations

Abbreviations are used to simplify the understanding of the cooling of transformers. Transformers are cooled using air (A), water (W), a mineral insulating oil (O), a synthetic nonflammable insulating liquid (L), natural convection (N), or forced convection (F). See Figure 5-1.

TRANSFORMER COOLING METHODS		
Transformer Type	**U.S. Abbreviation**	**Canadian Abbreviation**
Self-air-cooled (dry-type)	AA	AN
Oil-immersed self-cooled with air	OA	ONAN
Oil-immersed self-cooled with water	OW	ONWN
Oil-immersed forced-oil cooled with forced-water cooler	FOW	OFWN
Oil-immersed self-cooling with air-blast for additional cooling, permitting an increased rating	OA/FA	OFAF

Figure 5-1. Abbreviations are used to simplify the understanding of transformer cooling methods.

Self-Air-Cooled, Dry-Type

Self-air-cooled transformers are normally referred to as dry-type transformers. The coils and core are enclosed in a sheet metal enclosure that is provided with louvers or grates. The cooling of the coil windings is achieved by natural convection of the surrounding air and by the radiation of heat from the different parts of the transformer structure. See Figure 5-2. Self-air-cooled transformers have been in use for many years, but are restricted to small sizes because of heat problems. Modern technical developments of special high-temperature coil insulating materials that resist deterioration under heat have increased the size of these units to a 3750 kVA capacity at 15 kV.

Dry-type distribution transformers offer many advantages over liquid-filled transformers. Dry-type distribution transformers are ideally suited for indoor installation because they are safe and cannot explode. No toxic gasses are released and fire hazards are negligible. Some small transformers, such as bell-ringing, control, signaling, neon-sign, current instrument, and voltage instrument transformers are of the totally-enclosed type and the small amount of heat from the coil windings is dissipated by radiation to the surrounding air.

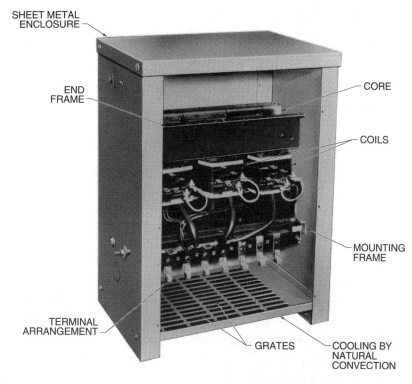

SHEET METAL
ENCLOSURE

END
FRAME

CORE

COILS

MOUNTING
FRAME

TERMINAL
ARRANGEMENT

GRATES

COOLING BY
NATURAL
CONVECTION

ABB Power T&D Company Inc.

Figure 5-2. Self-air-cooled transformers are cooled by natural convection of the surrounding air and by the radiation of heat from the different parts of the transformer structure.

Oil-Immersed Self-Cooling

In oil-immersed, self-cooling transformers, the complete core and coil assembly is enclosed in a metal tank and immersed in an insulating oil. The oil conducts the heat from the core and coil by convection to the surface of the enclosure where it is dissipated by radiation.

Plain Tank Wall. In small liquid-filled distribution transformers, the oil conducts the heat from the core and coil assembly by means of natural convection to the surface of the metal tank where the heat is then dissipated to the surrounding air. All types of liquid cooling depend on the motion (flow) of the warm liquid from the core and coil assembly upward and outward to the cooler surface of the tank. The liquid in self-cooled transformers rises when it is heated by the core and coil assembly and sinks when it comes in contact with the cooler tank surface. The temperature of the tank surface varies from top to bottom, depending on where the core and coils are located inside the tank. When the core and coils are located near the bottom of the tank, the tank as a whole is hotter, and more heat is dissipated at a given maximum temperature rise than when the core and coils are located near the top of the tank.

In small distribution transformers, such as pole-mounted distribution transformers commonly used in residential areas, the smooth tank wall used to encase the core and coil assembly has enough surface area to dissipate the generated heat by natural convection of the surrounding air and by radiation from the tank. See Figure 5-3.

The amount of iron in the transformer increases with the cube of its dimensions. The tank surface increases with the square of its dimensions. Therefore, as transformers are built larger and have increased kVA ratings, the amount of iron used in the construction of the core and the loss in the transformer increases at a faster rate than the physical dimensions of the tank surface. A point is reached where the tank surface cannot dissipate the heat. A means to increase the transformer's ability to do so must be applied. The heat-dissipating surface area of a tank can be increased by making the tank taller than is actually necessary to encase the transformer.

General Electric Co.

Figure 5-3. Small pole-mounted distribution transformers with oil-filled plain wall tanks dissipate heat by natural convection of the surrounding air and by radiation from the tank.

Tank Wall with Fins or Plates. When more heat must be dissipated than is permitted by a plain wall tank, additional radiating surface area can be obtained by welding radiating fins or plates to the walls of the tank. This increases the total surface area of the tank that is exposed to the surrounding air.

Tank Wall with Hollow Fins or Tubes. As the kilovolt-ampere rating increases, the length of time for cooling also increases. The heat generated in a transformer is proportional to the volume of material (product of the length, width, and height). The cooling ability of a unit is proportional to the external surface or to the square of the dimensions. As the kilovolt-ampere rating (or the volume) increases, the heat-dissipating ability of the unit also increases, but at a much slower rate. Therefore, it is necessary to provide additional means of cooling.

In medium-sized transformers, additional radiating surfaces are obtained by welding hollow radiating fins or tubes onto the tank. The hollow fins or tubes are welded to the top and bottom of the tank. See Figure 5-4. The fins are welded at the top and bottom of the tank by means of a nipple at each end which permits the oil in the tank to circulate by natural convection through the hollow fins or tubes.

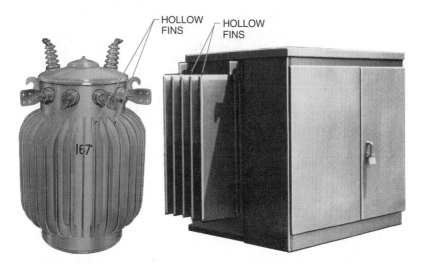

ABB Power T&D Company Inc.

Figure 5-4. In medium-sized transformers, additional radiating surfaces are obtained by welding hollow radiating fins or tubes to the tank.

Tank Wall with External Radiators. Liquid-immersed, self-cooled (OA) power transformers use external radiators for additional radiating surface. See Figure 5-5. The type and number of cooling units with a self-cooled (OA) transformer depends on the shipping clearance, rating of the unit, and other physical factors which are optimized at the time the transformer is designed.

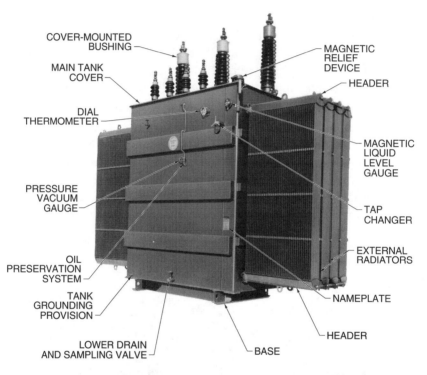

ABB Power T&D Company Inc.

Figure 5-5. Liquid-immersed, self-cooled power transformers use external radiators for additional radiating surface.

The external radiators consist of a group of hollow tubes which are connected to headers at the top and bottom. These headers are either welded to the tank wall or are attached by means of flanged connections through which the oil flows. The radiators for large units are normally connected to the tank by flanged connections so they may be removed for shipment. The radiators for small units are welded to the tank wall. In large units, shutoff valves are normally installed at the inlet and outlet connections so that, should any leaks develop, the radiator may be removed and repaired without removing all of the oil from the tank. See Figure 5-6.

Oil-Immersed Water-Cooling

Large self-cooled power transformers, due to their size, represent a considerable investment. Some operating companies prefer water-cooled transformers where water is cheap and available in sufficient quantities, and a low operating efficiency of the transformer is permissible. A copper cooling coil, wound in single or several parallel sections (depending on the quantity of tubing required), is placed in the upper part of the tank and supported from either the cover or from the tank wall. Water circulated through the coil(s) cools the

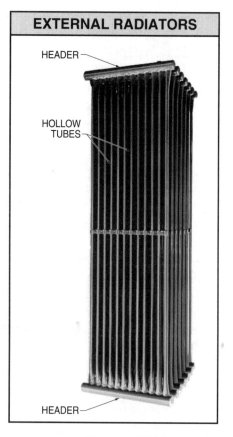

EXTERNAL RADIATORS

HEADER

HOLLOW TUBES

HEADER

ABB Power T&D Company Inc.

Figure 5-6. External radiators consist of a group of hollow tubes which are connected to headers at the top and bottom.

oil. The cooling coils are made of copper because iron coils are subject to corrosion. The cooling coils are tested with pressures as high as 500 psi as a safeguard against leaks while in service.

Oil-Immersed Forced-Oil Cooling

The one objection to water cooling is that if a leak occurs in the cooling coil, the water, which is at a higher pressure than the oil, enters the oil and greatly reduces its insulating quality. The water may reach certain live parts of the transformer and cause serious damage. A forced oil-cooled unit eliminates this hazard because the oil is circulated by pumps through an external water-cooled unit. The water pressure in the external cooling unit is less than the pressure in the oil circulating system. Any leaks cause the oil to enter the water of the external cooling system rather than water entering the oil. This type of cooling is used extensively in Europe and Canada but only occasionally in the U.S.

Oil-Immersed Self-Cooling with Air Blast

Additional heat may be dissipated with the addition of a blast of air against the radiators of large self-cooled units. The air blast permits increasing the rating of the transformer during the time that the air blast is supplied. The stream of air is provided by fans mounted on the radiators, or the radiators may be partially enclosed in a thin metal casing through which air is blown by means of a single large blower. The air-blast equipment is operated only when additional peak loads are required for short periods of time. The control for the equipment may be manual or automatic. See Figure 5-7.

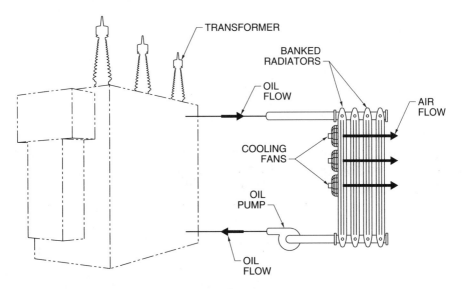

Figure 5-7. Additional heat may be dissipated with the addition of a blast of air against the radiators of large self-cooled units.

Automatically-controlled fans are started by a temperature sensor whose bulb extends into the hot oil. The temperature sensor contacts close, starting the fans when the oil reaches a predetermined temperature. The control equipment shuts OFF the fans when the oil temperature drops to a safe predetermined value. One complete control system is normally supplied with each transformer. In some cases, if the individual units are small, one control system is furnished per bank of transformers which are operated by a temperature sensor in the hot oil of one unit only. This arrangement simplifies and reduces the total cost of the control system equipment so that it can be used economically with smaller transformers. See Figure 5-8.

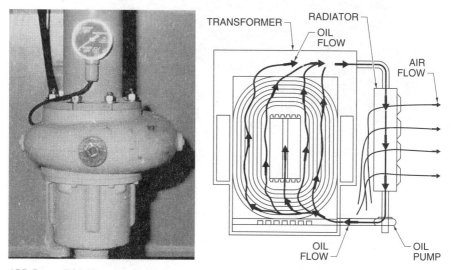

ABB Power T&D Company Inc.

Figure 5-8. An oil pump circulates oil through the transformer and radiator as fans force cooling air across the radiator.

Air-Blast Cooling

An air-blast transformer is a dry-type transformer used where great economy of space and weight is desired or where regulations prohibit the use of oil-insulated units because of the toxic fumes or fire hazard. Air-blast transformers are provided with air ducts between the coils, the coils and core, and the coil and core assembly and the metal housing so that air may be blown through them for the dissipation of heat. See Figure 5-9.

General Electric Co.

Figure 5-9. Air-blast transformers are provided with air ducts so air may be blown through the ducts for the dissipation of heat.

With the fans in operation, the transformers have increased ratings of 33% over the self-cooled ratings. Air-blast transformers normally function at 300 kVA to 15,000 kVA at a voltage not exceeding 15 kV. Transformer manufacturers are currently producing air-blast transformers with higher kVA and voltage ratings.

Heating and Temperature Control

The heating of a transformer is caused by the iron and copper losses. Immediately after a self-cooled transformer is put into service, the heat generated by the losses is stored in the iron and copper and their temperature increases. After continued operation, this heat is transferred to the oil and is carried by convection to the tank wall where it is radiated into the surrounding air. The temperature of the transformer continues to rise until the rate at which the tank wall dissipates the heat is equal to the rate at which it is generated. When this condition is reached for a given load, the transformer reaches a stable condition and operates continuously and indefinitely at this constant temperature and load, provided the maximum temperature reached is not high enough to damage the insulation.

Temperature Gradient

In an electrical circuit having resistance, a difference of potential causes a current to flow through the circuit. Similarly, a difference of temperature must exist between the windings and the oil to cause the heat to flow from the windings into the oil and to the tank wall. This temperature difference is the temperature gradient.

If a coil is relatively thick and has a large amount of insulation (which is not a good conductor of heat), the coil requires a large temperature gradient between the coil and oil to cause the heat to flow into the oil. If the coil is relatively thin with a small amount of insulation and large area exposed directly to the oil, the coil requires a small temperature gradient between the coil and oil to cause the heat to flow into the oil. This gradient depends on the coil construction. Furthermore, the temperature gradient varies as the total amount of heat generated in the windings varies, or as the copper loss varies (iron loss for a given voltage is constant). In terms of the load, the gradient varies approximately as the square of the load varies because the copper loss is proportional to the square of the load.

The temperature gradient is made up of two parts. The first part is from the point of maximum (hot spot) temperature to the average temperature of the entire winding. The second part is from the average temperature to the temperature of the hot oil. The hottest spot temperature is seldom known because it can be measured only by thermocouples or resistance coils embedded in the windings during manufacturing. This is rarely done because of the danger presented by the voltages of the windings. The average temperature can be determined by measuring the resistance of the transformer windings at a known temperature. For example, when the transformer is disconnected from the line and is at the same temperature as the surrounding air (ambient temperature), a resistance measurement is taken when the transformer is at load temperature. The temperature of the winding can be determined from these two resistances.

Temperature Limits

The temperature limit for a transformer is established by the materials which are used in the manufacturing of the transformer. The transformer insulation carbonizes under excessive heating because most of the insulation used in the windings consist of organic material and are fibrous in nature (such as kraft paper, manila paper, kraftboard, and pressboard). Continued high temperature accelerates the carbonizing action until the insulating strength of the transformer is greatly impaired. Chemists are constantly developing new synthetic materials with intermediate properties. One mixture that has been developed uses a special varnish pigmented with mica powder. This special mixture has a remarkable resistance to both oxidation and decomposition at temperatures well over 100°C.

There are many other types of insulating materials used in transformers. The effective properties of these materials largely depends on the structure in which they are used. Dielectric strengths, as measured by any of the standard tests, have relatively little relation to the dielectric stresses that can be used in a conventional transformer design. The National Electrical Manufacturers Association (NEMA) and the American National Standards Institute (ANSI) have established industry standards for dry-type transformers that stipulate the maximum temperature rise of the winding for various transformer insulating materials. The four insulating classes are Class A, Class B, Class F, and Class H.

Class A. Class A insulating materials are limited to use in transformers which are designed to have a continuous full-load temperature rise not exceeding 55°C over a 40°C ambient temperature. The standards allow for a hottest spot temperature of 10°C over the normal temperature rise. A Class A, 55°C rise transformer under full load has an average conductor temperature of 95°C when operating in a 40°C ambient temperature and a maximum conductor hottest spot temperature of 105°C. Class A insulating materials include:

- Cotton, silk, paper, and similar organic material when either impregnated or immersed in a liquid dielectric

- Molded and laminated materials with cellulose filler, phenolic resins, and other resins of similar properties

- Films and sheets of cellulose acetate and other cellulose derivatives of similar properties

- Organic varnishes (enamel) as applied to conductors

Class A insulating systems are normally used in oil-filled transformers. Class A transformer insulation is normally limited to small control transformers with ratings in the 1 kVA and below range because of the low temperature rise. They are also commonly used in large transformers where a high temperature is a hazard, such as in an area where combustible dust or fiber flyings may be present in the air.

Class B. Class B insulating materials are limited to use in transformers having a maximum temperature rise under full-load conditions not exceeding 80°C when the transformer is operating in a 40°C ambient temperature. The allowable hottest spot temperature can be 30°C higher. In other words, the average conductor temperature in a Class B 80°C rise transformer operating under full-load conditions in a 40°C ambient temperature is 120°C with the hottest spot in the winding being no more than 150°C.

Class B insulating materials consist primarily of materials or combinations of materials, such as mica and glass fiber, or other materials with suitable bonding substances capable of operating at an 800°C temperature rise. In the past, this was the most commonly used dry-type transformer. Even today, some manufacturers produce them through the 25 kVA rating range because of the lower inherent cost. Some individuals prefer the lower 80°C temperature rise because the transformer has a lower core loss, longer life, greater overload handling capabilities, and greater safety where low flash-point combustible flyings are present in the atmosphere. If the above advantages are required, use Class H insulated, 80°C rise transformers rather than Class B insulated transformers.

Class F. Class F insulating materials are limited to an average conductor temperature rise of 115°C under full-load conditions when operating in a 40°C ambient temperature. The allowable hottest spot temperature is 30°C higher. Therefore, when operating under full-load conditions in a

40°C ambient temperature, a Class F, 115°C transformer has an average conductor temperature of 155°C and a hottest spot temperature of 185°C.

Class F insulating materials consist primarily of materials or combinations of materials such as mica, glass fiber, or other organic and inorganic materials with suitable bonding substances capable of operating at a 115°C temperature rise. Other materials, both organic and inorganic, may be included in this class if by experience or accepted tests they are capable of operation at a 115°C average temperature rise.

The relative practicality of this insulation class for dry-type transformers is questionable. To legitimately qualify for this classification by NEMA and ANSI standards, the insulating materials must also qualify for the higher Class H temperature rises. The only difference is in the binding materials and impregnating materials. Many of the impregnating and binding materials that were once thought to be acceptable for 115°C operation have been found inadequate by NEMA and ANSI. Therefore, it appears that this insulation class will eventually fall into complete disuse as far as dry-type transformers are concerned. Class H insulating materials, which use inorganic bonding materials that are primarily composed of silicon compounds or resins, are used.

Class H. Class H insulating materials allow a design temperature rise of 150°C when a transformer is operating in a 40°C ambient temperature. This rise is the average rise found in the conductor and an additional 30°C hottest spot temperature is allowed. Therefore, the conductor temperature under full-load conditions when the transformer is operating at a 40°C ambient temperature reaches 190°C with the hottest spot not exceeding 220°C.

Class H insulating materials primarily consist of materials or combinations of materials such as mica, glass fiber, elastomer, and silicon, with suitable bonding substances such as appropriate silicon resins. Other materials or combinations of materials may be included in this class if, by experience or by accepted tests, they are capable of operation at 150°C temperature rise. These other materials must be present in very limited quantities for the entire unit to qualify as a Class H insulated transformer.

Most of the dry-type transformers manufactured today are of the Class H 150°C temperature rise classification. Standard dry-type transformers, 30 kVA and larger, are normally available from manufacturer stock. The high temperature rise of Class H insulation allows the transformer to be designed to a minimum size for convenience. Using a minimum amount of conductor and core permits it to occupy the smallest space and be the most economical of all dry-type transformers.

NEMA and ANSI standards stipulate that a transformer constructed with Class H insulating materials must withstand a 150°C temperature rise for 20,000 hours. If a transformer were designed to meet these exact specifications, it would have an expected life of under three years. This is unacceptable in most applications. To increase the life of the transformer, it is designed to operate at full-load well below the maximum 150°C temperature rise (normally in the 115°C range) and constructed with Class H insulation. This transformer has a long life and withstands long- and high-overload periods which results in low losses that normally offset the high purchase price within a relatively short time.

Ambient Temperature

The ambient temperature largely determines the load which can reasonably be carried in service. The actual temperature of a transformer is the sum of the ambient temperature and the temperature rise. For example, the actual temperature of a transformer with Class H insulating material (150°C) operating in a 40°C ambient temperature is 190°C.

Transformer Loading

Transformer loading is an important operating problem because most units do not have a steady load but operate on load cycles which may have a decided peak at certain hours of the day. The rated output of a transformer is fixed by its nameplate temperature rise measured under specified test conditions. The output, which it can deliver in service without causing undue deterioration of the insulation, may be more or less than the rated output, depending on the actual operating conditions.

A set of guides has been published by NEMA to guide the operator in safely loading oil-immersed power transformers. The guide covers general recommendations for loading all types of liquid-immersed power transformers specified to have an average winding temperature rise and hottest spot winding rise of not more than 65°C and 80°C respectively at rated load.

Transformers Equipped with Temperature Indicating Devices

Many transformers are equipped with protective and temperature indicating devices to protect the transformers against excessive temperatures and permit them to be loaded to their maximum safe operating temperature. On smaller distribution transformers, this device is actually a circuit breaker connected in series with the secondary windings and mounted inside the transformer case. See Figure 5-10. The circuit breaker is actuated by a bimetal strip comprising part of the breaker assembly through which the secondary load current flows. As the load current increases or decreases, the temperature of the bimetal strip increases or decreases. Also, as the oil temperature increases or decreases, the bimetal temperature increases or decreases by the same amount because it is mounted in the hot oil.

CIRCUIT BREAKER
IN SERIES WITH
SECONDARY WINDING

General Electric Co.

Figure 5-10. Small distribution transformers are equipped with circuit breakers to protect against excessive temperatures and permit loading to their maximum safe operating temperature.

The bimetal strip can be made to simulate the actual temperature existing in the transformer winding. When the bimetal reaches a temperature corresponding to the maximum safe operating temperature of the winding, any additional temperature causes the bimetal to release a latch and automatically trip the circuit breaker. This disconnects the load from the transformer and prevents injury to the windings. Some circuit breakers include auxiliary latches which close a circuit in which an indicating lamp is connected. The voltage for the auxiliary circuit is obtained from a small, low-voltage winding built in the transformer for this purpose. The auxiliary circuit latch trips and the indicating lamp gives a visible indication of the overload condition of the transformer when the transformer reaches its safe operating temperature. The indicating lamp is normally provided with a colored lens so that it can more readily be seen when ON.

The bimetal strip causes the main latch to release the tripping mechanism of the circuit breaker. The load is automatically disconnected if it and the temperature increase much above the safe operating temperature as indicated by the indicating lamp. As the unit cools, the circuit breaker may be closed by hand and service restored. Such a device permits maximum safe loading of the transformer at all times, regardless of the value of short-term peak loads.

Power transformers are not subject to fluctuating or large peak loads as are common on distribution transformers. For this reason, power transformers are normally equipped with devices which indicate only the temperature of the oil or the estimated hottest spot temperature of the winding. These devices are sometimes equipped with auxiliary contacts which close an alarm circuit actuated from an external source of voltage. The alarm circuit causes a bell to ring or a light to turn ON when the temperature of the oil or winding reaches a predetermined value.

The simplest device used for indicating the oil temperature, from which the temperature of the winding may be estimated, consists of an alcohol thermometer and an operating mechanism similar to a Bourdon gauge. See Figure 5-11. The alcohol thermometer is immersed in the hot oil and is connected to the operating mechanism by a flexible tube. An increase in oil temperature heats the alcohol in the bulb of the thermometer, causing it to expand. This exerts pressure which is transmitted through the flexible tube to the pointer of the operating mechanism. The pointer moves over a scale graduated in degrees. As the temperature decreases, the pressure decreases and the pointer moves to a low value on the scale. Some liquid temperature indicators are equipped with two pointers. One indicates the maximum temperature attained and the other follows the variations of temperature. This device may be designed to close contacts at a selected temperature. These contacts can be connected to start cooling fans or pumps, sound an alarm, or disconnect the transformer or load by operating primary or secondary circuit breakers.

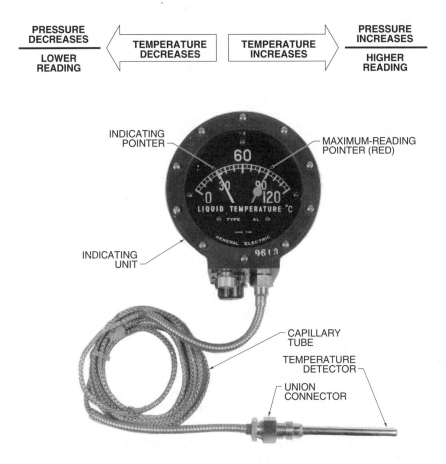

General Electric Co.

Figure 5-11. A liquid temperature indicator is used for indicating the oil temperature in a transformer.

A thermometer-type hottest-spot temperature indicator is normally used on large power transformers where the monetary loss is large due to damage caused by excessive temperature in the windings. This device gives the operator an indication of the temperature of the winding rather than that of the oil. One type of hottest-spot temperature indicator is an assembly which uses a Bourdon gauge connected to a temperature bulb by a capillary tube. The Bourdon gauge is calibrated in degrees Celsius. See Figure 5-12. The thermometer bulb and a heating coil are assembled in a well located in the hottest portion of the insulating oil near the top of the transformer. In most cases, the indicator dial is mounted on the transformer case, but can be placed a limited distance away from the transformer. The maximum distance is fixed by the length of the capillary tube connecting the indicator dial and the thermometer bulb.

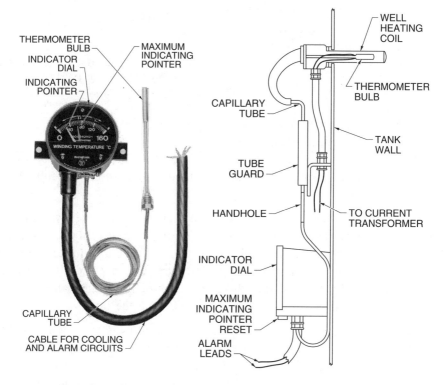

Westinghouse Electric Corp.

Figure 5-12. A thermometer-type hottest-spot temperature indicator is used on large power transformers.

The current in the winding of a power transformer passes through the primary winding of a current transformer, causing a current to flow in its secondary winding. See Figure 5-13. The current is proportional to that in the primary winding. The secondary current flows through a heating coil which encases the thermometer bulb, thus generating heat in the bulb proportional to that developed in the transformer winding. This increases the temperature to the thermometer bulb slightly, which is equal to the winding hottest spot rise above the hottest oil temperature.

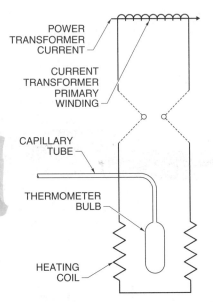

Figure 5-13. The current in the winding of a power transformer passes through the primary winding of a current transformer, causing a current to flow in its secondary winding.

The temperature of the bulb is at all times proportional to the temperature of the main transformer winding. Therefore, the instrument indicates the hottest spot temperature of the windings. Switches are normally provided in this instrument to operate forced-cooling controls and alarm circuits.

A bridge-type hottest-spot temperature indicator is another type of hottest-spot temperature indicator used only on power transformers. A bridge-type hottest-spot temperature indicator gives an indication of the temperature of the windings of a transformer. A thermometer is not used because the circuit is entirely electrical. For this reason, the indicator dial can be placed on a switchboard at any reasonable distance from the transformer.

A bridge-type hottest-spot temperature indicator operates when a current transformer is energized from the current in the main transformer winding. A heating coil is included in the secondary circuit of the current transformer. The amount of heat generated in the coil is proportional to the heat generated in the main transformer windings. A noninductive resistance embedded in the heating coil forms the fourth leg of a Wheatstone bridge. See Figure 5-14.

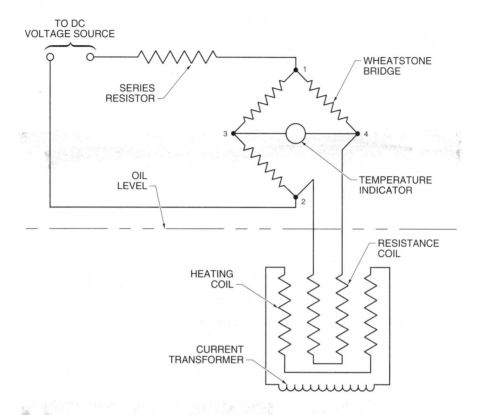

Figure 5-14. In a bridge-type hottest-spot temperature indicator, a noninductive resistance embedded in the heating coil forms the fourth leg of a Wheatstone bridge.

The Wheatstone bridge is completed by a series resistor outside of the transformer and is supplied voltage from a DC voltage source. A voltmeter is connected between two points, normally of equal potential.

When a current flows in the power transformer winding, a current proportional to the power current flows in the current transformer. The flow of current in the current transformer raises the temperature of the heating coil which increases the temperature of the noninductive resistance. The Wheatstone bridge becomes unbalanced and the amount of unbalance is indicated by the voltmeter. There is a definite relationship between the current and the variation of temperature. The voltmeter scale is calibrated to read the temperature of the heating coil directly in degrees Celsius. The Wheatstone bridge is balanced when normal current is flowing in the transformer. Any unbalance of the bridge is an indication of the temperature of the heating coil, which is proportional to the temperature of the main winding.

Transformer Oil

Oil is used in transformers to increase the strength of the insulating materials and to conduct the heat away from the coils and iron. Oil conducts heat to the outer tank of the transformer and its auxiliary cooling mediums, such as cooling tubes, coolers, and radiators, where it can be dissipated by radiation to the surrounding air. Refined mineral oils and nonflammable synthetic fluids (askarels) are used as transformer insulating and cooling liquids. The characteristics of insulating oil are similar to those of other petroleum oils. Askarels, such as chlorextol, inerteen, and pyranol, are composed of two synthetic chemicals with the addition of a fractional amount of a third compound.

When the oil of a transformer expands due to an increase in temperature, the air above the oil is forced out of the tank. Conversely, when the oil contracts due to decreased temperature, air is drawn into the tank. This breathing action is undesirable because the interchange of air brings oxygen and moisture from the atmosphere into contact with the oil. The moisture weakens the dielectric strength of the oil and the oxygen combines with the oil to form a sludge which darkens its color and causes deposits to form on the windings. These deposits may, in time, be sufficient to obstruct the oil circulation ducts placed in the windings. This results in temperatures higher than those for which the windings were originally designed. Ultimately, the insulation on the windings may become carbonized to such an extent as to cause a failure. See Figure 5-15.

To overcome the detrimental effect of breathed moisture, some transformers are equipped with dehydrating breathers. A *dehydrating breather* is a device containing calcium chloride which is connected to a pipe entering the tank above the oil level. Breathing takes place through the dehydrating breather and all moisture in the air is absorbed by the calcium chloride. A marker and scale indicate when the dehydrating agent has taken up a predetermined amount of moisture and should be replaced. Check valves are provided to prevent deterioration of the dehydrating material by contact with the atmosphere.

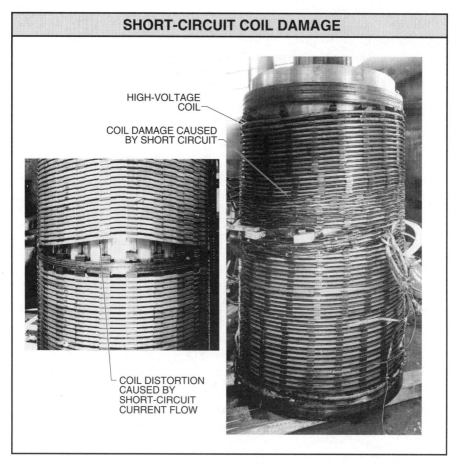

SHORT-CIRCUIT COIL DAMAGE

HIGH-VOLTAGE COIL

COIL DAMAGE CAUSED BY SHORT CIRCUIT

COIL DISTORTION CAUSED BY SHORT-CIRCUIT CURRENT FLOW

Kemper Insurance Co.

Figure 5-15. Increased coil temperature can lead to coil failure.

Some large distribution and power transformers are equipped with expansion tanks mounted near the top of the transformer tank to prevent the main body of oil from coming in contact with the oxygen in the atmosphere. See Figure 5-16. The expansion tank is connected by a pipe to the transformer tank, which is completely filled with oil. An increase in temperature, causes the oil to flow into the expansion tank which has a relatively small surface exposed to the air. The minimum oil level in the expansion tank is kept so that the transformer tank is always filled with oil. Breathing only occurs in the expansion tank and because it offers a much smaller area of oil surface exposed to the air than the transformer tank, the amount of oxygen absorbed is considerably less. The expansion tank may be provided with a dehydrating breather to absorb moisture contained in the air flowing in. The expansion tank is normally provided with a sump from which any accumulated moisture may be drained.

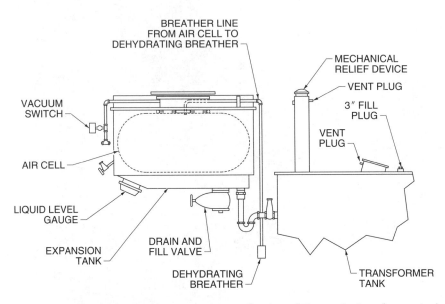

Figure 5-16. Expansion tanks mounted near the top of the main transformer tank prevent the main body of oil from coming in contact with the oxygen in the atmosphere.

The maintenance of oil-insulated transformers is a matter of preserving the qualities of the oil because the solid insulation of the transformer normally remains in good condition if the oil is properly cared for. The enemies of transformer oil are oxygen and moisture. Any device that operates to keep one or both from contact with the oil is a valuable addition to the transformer.

Some power transformers are equipped with Inertaire® equipment which effectively eliminates both moisture and oxygen from the transformer. See Figure 5-17. A transformer with Inertaire® equipment has the air above the oil space in the transformer tank replaced with pure dry nitrogen. The Inertaire® equipment consists of a supply tank of pure nitrogen connected to the space above the oil level at the top of the transformer tank. The supply of nitrogen is automatically controlled by a reducing valve and is conserved in the gas space by a mercury regulator. The mercury regulator permits the gas to escape if, due to a temperature increase, the pressure in the space above the oil exceeds a predetermined value. As the temperature of the oil decreases and the internal pressure falls below a predetermined value, the mercury regulator permits the release of additional gas from the nitrogen supply tank to the space above the oil.

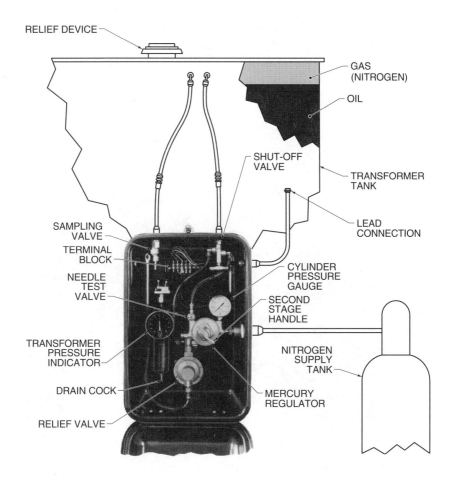

ABB Power T&D Company Inc.

Figure 5-17. Inertaire® equipment ensures long insulation life and negligible oil deterioration by maintaining a cushion of dry nitrogen above the transformer oil.

A transformer with Inertaire® equipment is protected against abnormal internal pressures which may develop due to a fault in the winding, although such faults are unusual. For this reason, a thin diaphragm of material such as micarta (which can be ruptured by such pressures) is used in combination with the manhole opening. The manhole cover holds the diaphragm in position. The diaphragm ruptures if an abnormal pressure develops inside the transformer tank. The impact of the pressure against the manhole cover raises it almost instantly and forms an annular space around its periphery through which the pressure is relieved. This cover is fastened to the diaphragm supporting ring by a set of heavy springs. See Figure 5-18. As soon as the pressure is relieved, the springs return the cover, thus preventing the escape of additional gas or the entrance of air and water into the gas space.

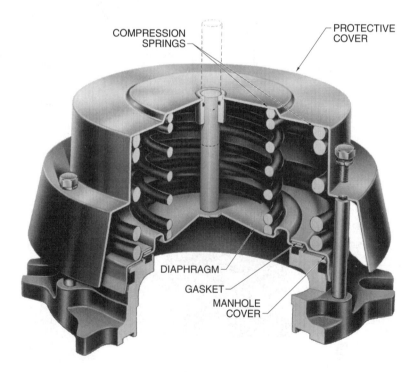

ABB Power T&D Company Inc.

Figure 5-18. An automatic resetting relief device is designed to relieve dangerous pressure which may build up within the transformer tank.

Mechanical Stresses

A magnetic field is produced in the region around a conductor whenever a current flows in the conductor. If another conductor also carrying a current is placed near the first, the resulting magnetic field at each point surrounding the two conductors is the vector sum of the fields. The field intensity is decreased in the space between the conductors and increased in the region outside if the currents are in the same direction. In this case, the two conductors are drawn toward each other. The field intensity is increased in the space between the two conductors when the currents are in opposite direction. The two conductors are forced farther apart.

When a current is flowing through a conductor which has been wound into a coil of several turns, the various turns attract each other at all points because the current is flowing in the same direction in all turns. When the primary and secondary coils of a transformer are assembled together, the currents in these two windings are opposite in direction. The force between them is a force of repulsion at all times. Under normal operating conditions, the mechanical stresses, which are developed due to these forces of repulsion, are very small. These stresses are multiplied many times during a short cicuit. If they are of sufficient magnitude, the coils are braced to prevent their injury due to the stresses when a short circuit occurs.

The core and coil assembly of a 3ϕ core-type transformer has the coils supported at both ends by support blocks with resilient pads. The support blocks with resilient pads are mounted on the frame which supports the entire core and coil assembly. The pressure is maintained against the frame to counteract the forces set up by heavy overload or short-circuit stresses. See Figure 5-19.

The core and coil assembly of a 1ϕ shell-type transformer has a permanently welded core clamp which cradles the core and minimizes stress. The magnetic circuit supports the coil against any displacement due to the short-circuit stresses on the portion of the winding enclosed by the magnetic circuit. See Figure 5-20. On large shell-type units wound with the pancake-type windings, the portions of the windings that extend beyond the magnetic circuit are braced with heavy steel frames and supporting plates.

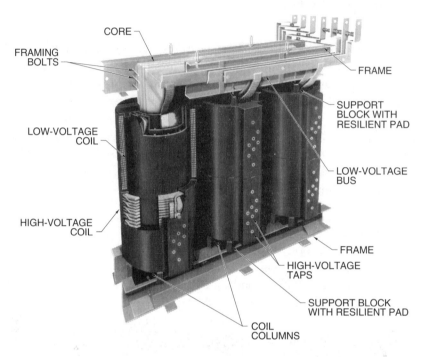

ABB Power T&D Company Inc.

Figure 5-19. A core and coil assembly of a 3ɸ core-type transformer has the coils supported at both ends by support blocks with resilient pads.

Under short-circuit conditions, the mechanical stresses depend on the frequency, size, voltage, and reactance of the transformer. Maximum stress is produced by a large transformer operated on a low-frequency circuit of moderate voltage and low impedance that is fed from a distribution sub-station of large capacity compared to the rating of the transformer.

Low reactance is desirable from the standpoint of good regulation. However, a transformer having low reactance may not be able to withstand the mechanical stresses set up during a short circuit.

Transformers must be protected from short circuits or grounds. The NFPA (National Fire Protection Association) No. 70 and ANSI C2 set specific requirements as to the rating of the required primary overcurrent protection. The rating of the required primary overcurrent protection device is smaller for dry-type transformers than for askarel- or oil-insulated transformers. The rating of the primary overcurrent protection device may be increased if the secondary is properly protected by an over-current protection device or is equipped with a coordinated thermal overload protection device by the manufacturer.

TRANSFORMER ACCESSORIES AND FITTINGS

The accessories found on power transformers enable safe operation of the transformer. The accessories found on power transformers include explosion vents, gas relays, level gauges, thermometers, bushings, and liquid-handling and sampling valves.

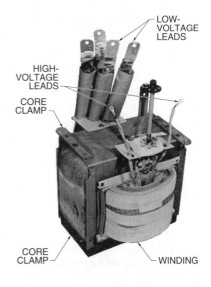

General Electric Co.

Figure 5-20. The core and coil assembly of a 1ϕ shell-type transformer has a permanently welded core clamp which cradles the core and minimizes stress.

Explosion Vents

High pressures are possible when an electrical fault occurs under oil. The pressure could burst the steel tank of a transformer. An *explosion vent* is a pipe, 4″ in diameter or greater, which extends a few feet above the cover of a transformer and is curved toward the ground at the outlet end of the pipe. Fitted at the curved end is a diaphragm that breaks at a relatively low pressure to release the forces from within the transformer. The diaphragm may be glass or thin phenolic sheeting (0.16″).

Gas Relays

The chemical and electrical phenomena associated with faults in oil-filled transformers generate gas. A considerable volume of gas is frequently formed in the early stages of a fault. For this reason, a protective device is needed to draw attention to the fault in its initial stage, thus allowing action to be taken to prevent serious damage.

The boiling range of transformer oil is 280°C to 400°C. At the upper end of the boiling range, traces of decomposition begin to appear. As the temperature rises higher, breakdown progressively increases. At arcing temperature, the oil is completely broken down into carbon and simple gases which are formed from the elements in the oil.

Electrical insulating oils decompose when subjected to an electric arc during silent discharge in regions of high electric stress, or when overheating occurs from high resistance joints, connections, or splices. The products of the decomposition are carbon, water or steam, carbon dioxide, and a number of flammable gases such as carbon monoxide, hydrogen, methane, and acetylene. The composition of gases produced varies with the oil, fault, power dissipated in the fault, and other factors.

The transformer must be removed from service as quickly as possible to prevent extensive damage, should such a fault develop. A thin diaphragm in the gas relay moves when acted upon by the pressure wave. A mercury switch connected to this diaphragm energizes relays to switch the transformer off the load. However, gradual overheating of any part, such as a hot joint, while not causing a pressure wave, can ultimately result in failure of the transformer. This local overheating decomposes or cracks the oil, forming gases which rise to the top of the tank. The gases are accumulated in a dome-like section of the relay in which a float is riding on the oil. The gas displaces the oil, dropping the level of the oil. Because the float rides the surface of the oil, the float also drops. When the float drops, it operates a mercury switch that engages an alarm circuit. Upon receiving an alarm, the condition may be investigated before extensive damage results.

Level Gauges

A level gauge is fitted to the transformer to ensure that the correct liquid level exists in the transformer. The level gauge is located on the transformer tank of small transformers that have no expansion tank. The level gauge is located on the expansion tank. A level gauge has a mark at the correct level for a 25°C temperature of oil. This ensures that should the temperature rise, the unit does not overflow nor can a dangerously low level be reached at low temperature conditions. Contacts are provided on large transformers to signal an alarm when the oil level drops to a dangerous level.

Thermometers

A thermometer is fitted to a transformer to measure the temperature of the top of the oil. In small transformers, this may be in the form of a liquid-filled thermometer mounted at the top of the tank. In large units, the thermometer is normally a gas-filled type that has the bulb fitted into a well in the cover. An indicator is located at the bottom of the tank at eye level. Contacts are provided to signal an alarm at high temperatures.

Bushings

The electrical power circuits must be insulated by bushings where they enter the transformer tank. In addition, the bushings and entrance must be oil-tight and weatherproof. The bushing is normally composed of an outer porcelain body. At high voltages, additional insulation in the form of oil and molded paper is used within the porcelain. The four types of bushings used on transformers as main lead entrances include solid porcelain, cable terminators or potheads, oil-filled, and condenser bushings.

Low-voltage transformers with separate leads normally have solid porcelain bushings. These bushings consist of high-grade porcelain cylinders through which the connections pass. The outside surface may be plain or have a series of corrugations or skirts to increase the surface leakage path to the metal case. When the conductors are brought to the transformers in lead covered cables, the leads often enter through cable terminators. These terminators are similar to the ordinary potheads and are attached to the transformer with the bushings inside the case.

High-voltage bushings are either oil-filled or condenser bushings. Oil-filled bushings have a central conducting rod or tube through which the conductor passes. Around this, a series of insulating barriers are held apart by spacers. The barriers and spacers are enclosed in a skirted porcelain shell which is filled with oil. Condenser bushings are similar except that the central rod is wound with alternating layers of insulation and tinfoil. This results in a path from the conductor to the case consisting of a series of condensers. The layers are designed to provide an approximately equal voltage drop between each condenser. In some condenser bushings, the whole bushing is enclosed in a skirted porcelain shell. Other types have only the exposed part of the bushing enclosed in a skirted porcelain shell. See Figure 5-21.

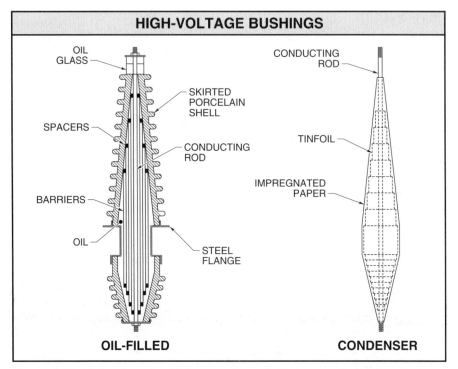

HIGH-VOLTAGE BUSHINGS

OIL
GLASS

SKIRTED
PORCELAIN
SHELL

SPACERS

CONDUCTING
ROD

BARRIERS

OIL

STEEL
FLANGE

OIL-FILLED

CONDUCTING
ROD

TINFOIL

IMPREGNATED
PAPER

CONDENSER

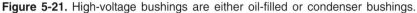

Figure 5-21. High-voltage bushings are either oil-filled or condenser bushings.

Liquid-Handling and Sampling Valves

A liquid-handling valve is fitted at the bottom of the transformer tank to add or remove liquid from the tank. The valve is normally connected to a sump to ensure that all liquid is removed. Adjacent to the liquid-handling valve is a ½″ sampling valve which also leads to the sump. A *sampling valve* is a needle valve which is used to draw a sample of liquid for test purposes. By taking liquid from the lowest point in the tank, any free water should appear in the sample and give warning of contamination of the liquid.

Transformer Nameplates

A transformer, like all other electrical equipment, is designed and manufactured to function in an electrical system at a specified voltage, frequency, load, etc. The nameplate data is carefully checked to ensure that the transformer is installed and maintained according to the data. The

nameplate should never be defaced
or removed. See Figure 5-22. The
information contained on the nameplate of a transformer is normally
divided into five sections. The five
sections include:

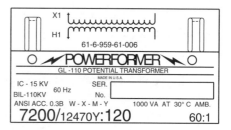

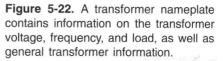

- General – Information on
  transformer serial number,
  type, rating, nominal voltages,
  impedance, tap changer range,
  oil capacity, weights, etc.

Figure 5-22. A transformer nameplate
contains information on the transformer
voltage, frequency, and load, as well as
general transformer information.

- **Physical Terminal Arrangement** – Normally as a block drawing, the
  physical locations of all terminals are given with reference to some
  obvious feature of outside construction. This enables all connections
  to be positively identified.

- Schematic Diagram – A complete winding schematic diagram shows
  all internal connections, coil tap numbers, internal selector switches,
  and main terminal markings. The combined use of the physical terminal arrangement drawing and the schematic diagram enables the
  user to determine exactly how each bushing is connected internally.
  Also included on the schematic diagram are all auxiliary components
  such as current and potential transformers showing their actual electrical positions and the ratios.

- Vector Reference – On all 3ϕ units, primary and secondary voltage
  reactors are drawn to indicate the phase angle between the primary
  and secondary voltages. The vector diagrams are labeled with the
  terminal designations used on the schematic diagram.

- Voltage Ratio Table – One or more tables are given for transformers
  with a tap changer. The table lists the tap indicator number, the
  actual coil numbers, and how they are connected. The voltage ratio
  is given for each tap position and, in some instances, the full-load
  current at that tap position. On many nameplates, to avoid the use
  of voltage ratios, which would mean small decimal numbers, the
  nominal voltage of the winding without the tap is taken and the
  voltage required to produce this nominal value is listed for each
  tap. Both methods give the same information.

Impedance. To determine the impedance that appears on the nameplate
of a transformer, apply the procedure:

1. Connect an ammeter across (in series) the secondary winding.

2. Connect a variable rheostat in the primary winding circuit.

3. Connect a voltmeter across the primary winding.

4. Apply a voltage to the primary circuit. Adjust the rheostat to increase the voltage across the primary winding until the secondary winding current reaches its rated current.

5. Read the voltage across the primary winding.

6. Divide the applied primary winding voltage by the rated primary winding voltage.

For example, a transformer has a 2400 V, 10 A primary rating and a 240 V, 100 A secondary rating. A voltage of 120 V is needed across the primary winding for the ammeter in the secondary winding circuit to read 100 A. Dividing 120 by 2400 equals .05 or 5% impedance. See Figure 5-23.

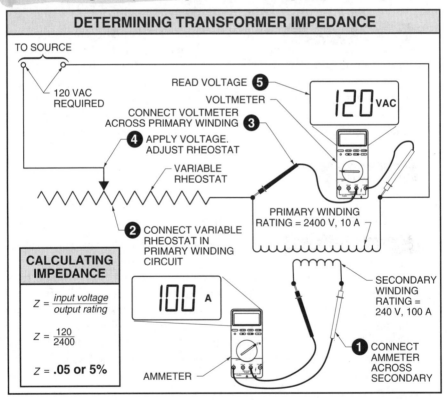

Figure 5-23. The impedance of a transformer is determined by connecting a variable rheostat, voltmeter, and ammeter in the primary and secondary windings.

Transformer Sound Levels

All transformers hum when energized. The hum is due to the vibrations generated within the laminated steel core structure. The hum has a fundamental frequency of twice the applied frequency. The volume is determined by the transformer design, construction characteristics, and the methods used in their installation. The level of sound produced by a transformer is measured in decibels (db). A *decibel* (db) is the unit used to measure the intensity level of sound. It is equal to $\frac{1}{10}$ of a bel (unit named after Alexander Graham Bell). The decibel number represents a ratio of the level of sound to a reference level (usually 0 db). For every 10 db change, the sound level increases by 2 times or decreases by $\frac{1}{2}$ times. For example, the sound level of normal conversation is one-half of a hair dryer. Test procedures have been established so that transformer manufacturers can publish the sound level ratings of their transformers. Sound levels are based on the kVA rating of the transformer. See Figure 5-24.

TRANSFORMER SOUND LEVELS			
kVA Rating	Sound Level (in db)	Loudness	Example
0–5	40	Moderately quiet	Refrigerator running
6–9	40		
10–25	45		
26–50	45		
51–150	50	Quiet	Whisper
151–225	55		
226–300	55		
301–500	60	Moderate	Normal conversation

Figure 5-24. Transformer sound levels are based on the kVA rating of the transformer.

The sound level produced by a transformer is normally not a problem when transformers are installed in substations, vaults, or outdoors. There are specific critical locations where sound is an important factor. The noises in and around most indoor locations, (ambient sound level) normally mask transformer sounds. Tables are available for the average ambient sound levels in areas where transformer noise could be a problem. See Figure 5-25.

AMBIENT SOUND LEVELS	
Area	Sound Level (in db)
Residences (including apartment buildings)	25–45
Retail stores	45–60
Office areas	45–75
Manufacturing plants	75–95

Figure 5-25. Tables are available for the average ambient sound levels in areas where transformer noise could be a problem.

For example, if a 150 kVA distribution transformer with a 50 db rating is installed in a factory that has an ambient sound level of 85 db, the sound of the transformer would not be heard above the ambient sound. If this same transformer were installed in an apartment building where the ambient sound level is 30 db, the transformer sound would be noticeable and would be considered objectionable. In other areas such as schools, churches, and hospitals where the ambient sound level is very low, special precautions must be taken to select a transformer with a low sound rating and to locate and install the transformer to keep the sound level at a minimum.

Low-Sound Level Installations

When two transformers with equal sound levels are installed side-by-side, their combined sound level is only 3 db higher than their individual level. The combined level of three similar transformers is 4.8 db higher than their individual level. For example, if three 50 kVA transformers with individual ratings of 45 db are installed in the same area, their combined sound level is 49.8 db (45 db + 4.8 db = 49.8 db), not 135 db. Although this could be an important factor, other installation precautions must be taken when installing transformers if the sound level is to be kept at a minimum. These precautions include:

- Selecting a transformer that has a low-sound level rating. The sound level rating of the transformer should be below the ambient sound level of the area where it is to be installed. In some critical, low-sound level areas, it may be necessary to factory order a special quiet-type transformer. Such an area could be in a hospital, where the NEC® requires that isolating transformers be installed to feed special circuits in operating rooms.

- Locating the transformer as far away as possible from areas where sound is objectionable.

- Locating the transformer where the sound level is not amplified by sound reflection from the walls or ceiling. Tests have shown that the least desirable location for a transformer is in a corner near the ceiling because the walls and ceiling amplify the sound. If the transformer is free-standing and the room where it is to be installed is near a low-sound level area, the only accurate way to determine the lowest sound level location is to make a temporary connection and move the transformer to different locations in the room. The walls and ceiling of the room could be covered with acoustical tile or fiberglass which dampens the sound of the high harmonics of the transformer but have little effect on the fundamental hum generated by the transformer.

- Mounting the transformer so that the mechanical vibrations of the transformer are not transmitted to the connected raceway system or to the structural parts of the building.

Rigid mounting on a heavy reinforced concrete wall or floor is normally suitable for small transformers because their mass is comparatively small. When mounting transformers on a structural frame, wall, ceiling, or column, a flexible mounting is normally necessary. A flexible mounting requires the transformer to be installed on flexible mounts which isolate the transformer vibrations from the structural part of the building. These mounts must be properly selected and loaded. They are normally furnished and installed by the transformer installer. Mounts must be installed so there is no metal-to-metal contact between the transformer and the structural part of the building. On specially constructed low-sound level transformers, internal vibration pads, which isolate the core and coil assembly from the enclosure, are installed at the factory by the manufacturer. When small transformers not having built-in vibration isolators are installed, the conduits or wiring method must be flexible to eliminate the conducting of the transformer vibrations to the structural part of the building.

Transformer Frequency

In the U.S., the standard frequencies in common use are 25 Hz and 60 Hz. See Figure 5-26. The frequency of 60 Hz is used for lighting and power installations, while 25 Hz is often used for railroad electrification. A frequency of 125 Hz has been used for aircraft electrification. In Europe, frequencies of 50 Hz and 16⅔ Hz are commonly used. 50 Hz is used for lighting and power installations and 16⅔ Hz is used for railroad electrification.

STANDARD FREQUENCIES		
U.S.	**Europe**	**Use**
25 Hz	16⅔ Hz	Railroad electrification
60 Hz	50 Hz	Lighting and power

Figure 5-26. The standard frequencies in use in the U.S. is 25 Hz and 60 Hz.

The 25 Hz electricity is not used for lighting and power systems because the 25 Hz system would have a lower impedance due to the lower inductive reactance. This type of system would also have a better voltage regulation. There are other advantages of the 25 Hz system, but the basic problem is that when this system is used for lighting, the filament of the lamp cools down between cycles at this lower frequency. This causes a noticeable flickering of the lamps. At the frequency of 60 Hz, the filament of the lamp is subjected to the low current for a much shorter time and the filament does not have time to cool, therefore the flickering is not noticeable.

When selecting or installing a transformer, the transformer must be rated to operate at the same frequency as the electrical system. The altitude at which power transformers operate must also be considered. Altitudes above 3300′ (1000 m) may require kVA correction for temperature rise at high altitudes. While high altitudes do not affect the size of the core or coils, the transformer may require an increase in radiating surface.

The current of a transformer varies inversely to the frequency if the voltage applied to the transformer is kept constant. As the inductance increases with the number of turns, the number of turns in the primary may be decreased as the frequency of the primary voltage is increased. If a 60 Hz transformer is connected to a 25 Hz system, the current of the primary becomes more than twice the current rating of the primary. For this reason, serious damage could be done to the primary winding of the transformer.

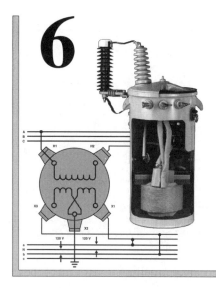

6

TAP CHANGERS

Transformers are provided with tap changers to provide a uniform secondary voltage when the primary voltage varies due to line voltage drop. Tap changers are used to change taps with the transformer de-energized. The parallel-winding and single-winding methods are the two methods of changing taps on a transformer without interrupting the load.

TAP CHANGERS

A *tap* is a connection brought out of a winding at some point between its limits to allow changing the voltage or current ratio. A *tap changer* is a selector switch device used to change transformer taps with the transformer de-energized. Tap changers may be found in one or both windings of a transformer. The taps are adjusted manually in distribution transformers. Tap changers are adjusted automatically or by some external means, such as supervisory control from a central dispatch location on power transformers used in the transmission of electrical power.

Tap changers are also used with autotransformers to form a regulator or booster for varying the voltage in small steps on distribution feeder lines. They are also used for adjusting power interchange and for phase-angle shifting on interconnected transmission systems.

Transformer taps allow a uniform secondary voltage to be provided when the primary voltage varies due to line voltage drop. See Figure 6-1. The $4\frac{1}{2}\%$, 9%, and $13\frac{1}{2}\%$ taps cut out the corresponding percentage of the primary winding. This changes the ratio of primary turns to secondary turns.

The primary voltage setting is related to the constant secondary voltage of 120 V or 240 V. See Figure 6-2. For example, the transformer must be put on the $4\frac{1}{2}$% tap (full winding − $4\frac{1}{2}$%) to maintain the secondary voltage at 120 V or 240 V if the primary line voltage drops to approximately 2292 V.

Taps on a transformer may be changed by considering the voltage at peak load and off-peak load. Voltage swing is determined by placing a recording voltmeter in the secondary circuit for at least 24 hours to determine the lowest and highest voltage. A tap set for peak load voltage has too high a voltage during off-peak load. A satisfactory voltage is required at all times. A satisfactory voltage is a voltage between 117 V and 122 V at the customer's service entrance. See Figure 6-3.

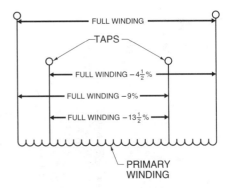

Figure 6-1. Transformer taps allow a uniform secondary voltage to be provided when the primary voltage varies due to line voltage drop.

TRANSFORMER WINDING RATIOS

Tap Winding	Primary Voltage*	Secondary Voltage*	Ratio
Full Winding	2400	120 (240)	20:1 (10:1)
$4\frac{1}{2}$%	2292	120 (240)	19:1 (9.5:1)
9%	2184	120 (240)	18:1 (9:1)
$13\frac{1}{2}$%	2076	120 (240)	17:1 (8.5:1)

* in V

Figure 6-2. The primary voltage setting is related to the constant secondary voltage of 120 V or 240 V.

SATISFACTORY VOLTAGE LEVELS

Devices	Tolerable Low Voltage	Satisfactory Voltage	Tolerable High Voltage
Lights, motors, etc.	112	117	122–127
Stoves, motors, etc.	230	235	245–250

Figure 6-3. A satisfactory voltage is a voltage between 117 V and 122 V at the customer's service entrance.

The secondary voltage and winding ratio of a transformer are required to determine the correct primary tap setting of the transformer. For example, in a 2400 – 120/240 V transformer, the secondary voltage is measured at 226 V with the tap setting on full winding (2400 V). The secondary voltage should be 240 V. The ratio of the windings is calculated by dividing the primary voltage by the secondary voltage. The ratio of the windings is 10:1 (2400 ÷ 240 = 10 ÷ 1 = 10:1). The primary voltage is calculated by multiplying the ratio of the windings by the secondary voltage. The primary voltage is 2260 V (10 × 226 = 2260 V). By putting the tap setting on 4$\frac{1}{2}$% (changing the ratio), a voltage of 238 V is obtained. 238 V is a satisfactory voltage level for the required 240 V.

No-Load Tap Changer

Taps may be brought to a terminal board inside the transformer tank if the change in voltage ratio is infrequent. See Figure 6-4. Taps are changed after the transformer is disconnected from the line by changing the position of one or more terminal strips which connect the studs on the terminal board. The taps from the winding are connected to these studs. Connections are made according to instructions found on the nameplate or diagram of connections supplied by the transformer manufacturer.

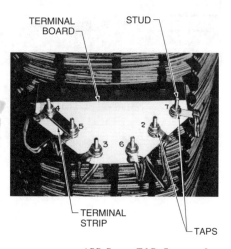

ABB Power T&D Company Inc.

Figure 6-4. Taps may be brought to a terminal board inside the transformer tank if the change in voltage ratio is infrequent.

This method of changing taps is inconvenient because it requires the removal of the main cover or handhole cover and working, at times, in hot oil. The operator may accidentally drop a tool or loose metal part, such as a washer or nut, from the terminal board into the transformer winding. A careless operator may also make the wrong connection and cause serious trouble within the transformer or to the apparatus connected to it. This method of tap changing is used for distribution and small power transformers because it is the cheapest method of tap changing from the standpoint of initial cost.

Manually-Operated Switch

The taps from the winding may be brought to a switch instead of a terminal board having studs and connectors. This safeguards the transformer from improper connections and decreases the time required to make a change of taps. This switch is operated by a handwheel or handle. This handwheel or handle may be on the underside of the cover for internal operation or may be connected to a shaft extending through the cover or side of the tank wall for external operation.

In either case, the transformer must be disconnected from the line before the change of taps is made. Such a method of changing taps eliminates the possibility of making the wrong connections, dropping loose parts into the winding, or having high resistance in contacts due to improper tightening of connectors to studs on the terminal board. This also eliminates the necessity of having to remove the cover when the handwheel is on the outside of the transformer tank. See Figure 6-5.

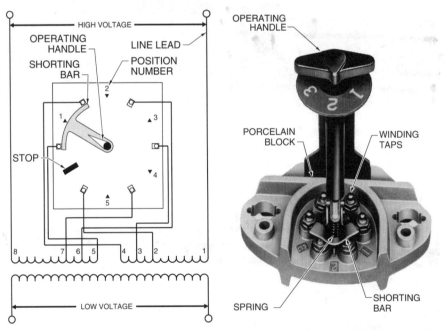

ABB Power T&D Company Inc.

Figure 6-5. The taps from the winding may be brought to a switch to safeguard the transformer from improper connections and decrease the time required to make a change of taps.

In position 1, the full high-voltage winding is connected to the line. As the insulated handle is rotated to the right (clockwise) to positions 2, 3, 4, and 5, alternate left and right sections of the winding are removed from the circuit. In position 1, the winding taps 4 and 5 are connected together by a current-carrying shorting bar. The shorting bar is large enough to span two adjacent studs to which the taps are connected. The shorting bar is part of the operating handle but is insulated from it. As the handle is turned to position 3, the shorting bar connects winding taps 3 and 6, omitting all turns between taps 3 and 4 and between taps 5 and 6 from the circuit.

Tap Changing Under Load

With the expansion and interconnection of power systems, it is often necessary to change the transformer taps several times daily to obtain the required voltage variation. The demand for continuity of service and the inconvenience of having to disconnect the transformer frequently from the line to change the voltage ratio have led to the use of equipment that enable the taps to be changed while the transformer is connected to the line and carrying a load. Such equipment for changing transformer taps under load is particularly adaptable where voltage control is required in connection with large blocks of power. The parallel-winding and single-winding methods are the two methods of changing taps on a transformer without interrupting the load.

Parallel-Winding Method. In the parallel-winding method, one winding of the transformer is arranged with parallel circuits. See Figure 6-6. Each of the parallel windings is provided with taps which are changed only when their respective winding is opened by a circuit breaker. In normal operation, the two transformer windings are operated in parallel, therefore each winding is only half of the capacity of the third winding. However, each of the parallel windings must be capable of carrying the full-load current of the transformer while the taps are being changed on the winding which is opened.

Taps 2 to 9 of parallel windings A and B are each brought to an externally-operated tap changer which is controlled through a system of cams and gears so that the proper sequence of tap changing is obtained when operated. This precaution is necessary to ensure the same amount of winding is cut out of each of the two sections, otherwise a heavy circulating current would flow in the windings when they are connected in parallel.

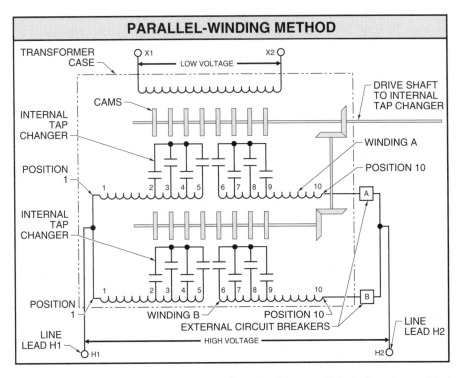

Figure 6-6. In the parallel-winding method, each of the parallel windings is provided with taps which are changed only when their respective winding is opened by a circuit breaker.

The ends of the two windings (position 1), are connected directly to outgoing line lead H1. The leads at position 10 are connected to external circuit breakers A and B, which are connected in parallel to the outgoing side of line lead H2. The internal tap changer mechanism is operated through a driveshaft extending through the transformer case.

The voltage ratio of a parallel-winding transformer is changed by opening circuit breaker A, changing taps in winding A, closing circuit breaker A, opening circuit breaker B, changing taps in winding B (same as in winding A), and closing circuit breaker B. This procedure is repeated for each change of taps.

As circuit breaker A is opened, winding A is disconnected from the line and the complete load is carried by winding B. When circuit breaker A is closed and B is not yet opened, the load is divided between windings A and B. As circuit breaker B is opened, the load is transferred to winding A. Closing circuit breaker B, causes windings A and B to each carry half of the load. This procedure for tap changing has been replaced to a large extent by the single-winding method.

Single-Winding Method. In the single-winding method of tap changing under load, a reactor or preventive autotransformer is used to bridge across the taps on the main transformer winding. The use of an autotransformer also gives an additional voltage step between each adjacent tap. For example, if the transformer has four, 5% taps, an additional $2\frac{1}{2}\%$ step is obtained during each tap-changing cycle when the autotransformer is connected across two adjacent taps. This is possible because one end of the main transformer winding is connected to the midpoint of the autotransformer. See Figure 6-7.

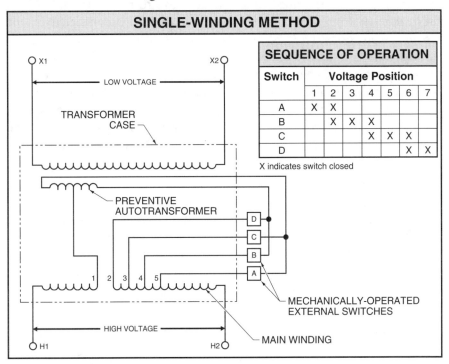

Figure 6-7. In the single-winding method, a reactor or preventive autotransformer is used to bridge across the taps on the main transformer winding.

With the autotransformer connected in position 1 with only switch A closed, the transformer is connected across-the-line with all the taps out of the circuit and half of the autotransformer winding in series with the main winding. This portion of the autotransformer carries all the load current. In position 2, with switches A and B closed, the autotransformer is connected across taps 4 and 5, and half of the load current flows through each half of the autotransformer winding. The voltage across H1 and H2 is increased by one-half of the voltage between taps 4 and 5. In position 3, with only switch B closed, half of the autotransformer winding is in series with the main winding, but the voltage has been stepped up across H1 and H2 by an amount equal to the voltage between taps 4 and 5.

This sequence of operation can be continued until all the tap sections of the main transformer are inserted in the circuit. One or more of the breakers are always closed, therefore the variations of voltage ratio are obtained without interrupting the load connected to the transformer.

Various modifications of this circuit may be employed. For example, in the circuit just described, the voltage taps are not equal because of the reactance drop of the autotransformer when it is connected to one tap only (positions 1, 3, 5, and 7). When full-load current is passed through one-half of the autotransformer with the other half open, the full-load current of the main transformer becomes the exciting current of the autotransformer. Under this condition, there are no neutralizing ampere turns from the other half, so the autotransformer becomes a reactor. Air gaps are provided in the core to prevent saturation and to give low impedance when operating in this manner. This makes the exciting current relatively high when the autotransformer is connected across taps.

A circuit breaker is sometimes added to this circuit to make the voltage steps exactly equal for each position. After the switching cycles are completed to positions 1, 3, 5, and 7, the circuit breaker short circuits the autotransformer, removing it from the circuit. This circuit breaker is mechanically interlocked with the externally-operated switches so that the alternate positions 2, 4, and 6 cannot be made without first opening the circuit breaker, again placing the autotransformer in the circuit during the switching operation.

Tap Changer Locations

There is no fixed rule for the location of the tap changer with respect to primary or secondary windings. In general, it is desirable to locate the tap changer in the high-voltage winding because of its smaller current. Its position may be fixed by the application of the transformer, which may require parallel operation on one winding or the regulation of power flowing in a given direction. If the voltage of the high-voltage winding becomes so great as to make it difficult to insulate the tap-changer parts, the tap changer is sometimes put in the low-voltage winding. If the current in the low-voltage winding is too large for the capacity of the tap-changing switch, or the voltage of either winding is too high, a series transformer is required.

In a wye-connected transformer with the neutral solidly grounded, the taps and tap-changing switches are placed in the grounded neutral end of the winding. Although the winding may have a very high voltage at its line end, the maximum voltage to which the tap changer switches are subjected is equal only to the voltage existing between the neutral and the highest tap position. The nearer the taps are placed to the neutral end, the lower the voltage.

Automatic Control. A motor-driven tap changer is used to compensate for voltage drops in applications where the variation in voltage is considerable throughout the day.

If the transformer is unattended, voltage drop control may be achieved by pulsing the tap changer's drive motor with a voltage regulating relay. Operation of this relay causes the motor to rotate in such a direction that taps are changed which maintain the voltage at nearly a constant value. As some voltages are self-correcting, a time-delay relay is used in conjunction with a voltage regulating relay. This allows a certain length of time to elapse between the moment the voltage variation is picked up and the motor is set in operation. This device eliminates a large number of unnecessary operations and greatly increases the life of the apparatus.

Remote Manual Control. A remote hand switch, or supervisory control, is used in applications where it is necessary to correct for voltage drop or phase angle only during peak load periods. Both of these devices actuate the tap changer's drive motor from a remote location and are used in place of the voltage regulation relay.

Hand Crank. For maintenance or emergencies, a hand crank is provided so that the mechanism can be operated at the transformer. The crank is usually the means by which a safety switch in the motor circuit is closed so that the motor cannot be operated while the crank is being used.

Phase-Angle Control

It is necessary in some cases to transmit power to a given load center over two parallel transmission lines. This is due to the rapid increase in the use of electricity and the demand by consumers for uninterrupted service. Also, to generate the electricity more economically, it is advisable to interconnect several generating stations so that, as the load becomes too large for one station, additional stations or generators may be placed on the line. As the load center changes, the load may be shifted from one station to another. To do this, it is necessary to regulate the phase displacement between the two parts of the circuit while carrying the load. This phase-angle control is obtained by adding a suitable voltage to the line at some angle with the line voltage.

For example, two systems that are to supply equal loads to a given load center are of unequal length, thus have unequal impedances. See Figure 6-8. Point B represents the load center with load supplied over lines C and D connected in parallel. IR_C and IX_C represent the resistance and reactance drops in line C due to the load of I amps in line C. If a voltage E_B is required at the load center, adding IR_C and IX_C vectorially to E_B gives the voltage (E_C) required at point A to send the load of I amps to point B.

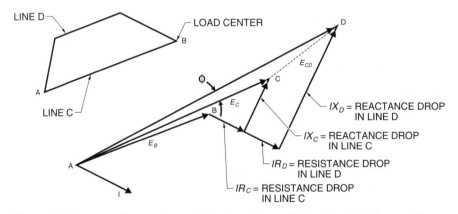

Figure 6-8. A vector diagram is used to show phase displacement between voltages of two lines of different impedances supplying equal loads.

If line D has a greater impedance than line C, and must carry the same amount of current, the resistance drop (IR_D) and the reactance drop (IX_D) are greater than in line C. The resultant voltage (E_D) required at point A to send the current to point B at a fixed voltage of E_B, is E_B, IR_D, and IX_D added vectorially. The voltage E_D required at point A is greater than the voltage E_C required at point A and the two voltages are out-of-phase with each other.

If each line is to carry an equal power load and equal reactive power when connected together, it is necessary to either increase the impedance of line C to that of line D so that the impedance drop in both lines are the same, or add a voltage (E_{CD}) to the circuit. E_{CD} is the difference in voltage between E_C and E_D. Because the voltages E_C and E_D are not in phase with each other, it is not sufficient to add voltage to line C by changing the voltage ratio. A voltage must be added which is at some angle with the reference voltage.

A vector diagram is used to show the principles of phase-angle control. See Figure 6-9. E_B represents vectorially the reference voltage. $+E\phi$ and $-E\phi$ are the voltages that must be added at right angles to voltage E_B to obtain a resultant voltage E_R which has a phase displacement of ϕ degrees with respect to the reference voltage E_B. Voltage control is obtained by varying the length of the reference voltage vector E_B.

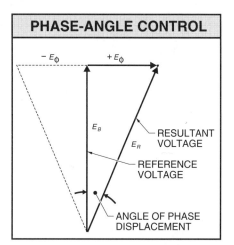

Figure 6-9. A vector diagram shows the principles of phase-angle control.

The angle ϕ, which represents the angle of phase displacement, can be changed by varying the length of the voltage vector E which is a quadrature voltage with respect to the voltage E_B. The voltage represented by the vector E_B can be controlled by changing the voltage ratio of the transformer by means of the tap changing-under-load equipment. The quadrature voltage vector E can be controlled similarly by combining suitable variable quadrature components of voltage with the normal line voltage.

Phase-angle control may be accomplished using a 3φ regulator unit. See Figure 6-10. Assume that approximately ±6° of phase control is desired in eight 1½% steps. The equipment consists of an exciting transformer excited from the line which is to be controlled, with its secondary connected to a series transformer inserted in the same line.

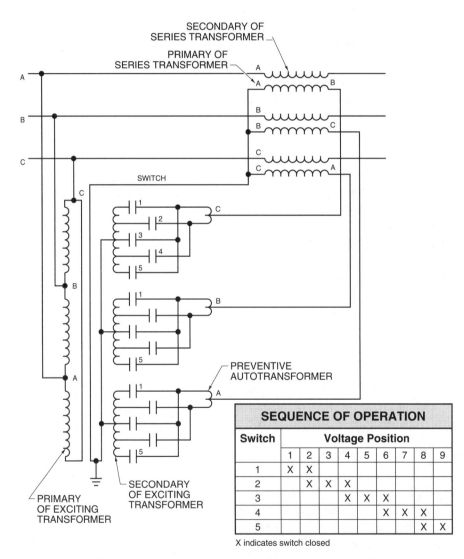

Figure 6-10. Phase-angle control may be accomplished using a 3φ regulator unit.

Each secondary winding of the exciting transformer is divided into four equal parts so that, with the use of a preventive autotransformer, eight steps in voltage are obtained. The principle of operation of the regulating transformer secondary and autotransformer is the same as that of the single-winding method of voltage control. The same type of equipment is used for both.

The objective of the equipment is to obtain a voltage in the secondary of the series transformer, which is at right angles to the normal line-to-neutral voltages AN, BN, and CN. See Figure 6-11. The primary of the exciting transformer is connected in delta. The secondary is connected in wye, with the neutral made in the middle of each winding. With the neutral so connected, the voltage impressed across the primary of the series transformer varies; from a maximum in the positive direction through zero, to a maximum in the negative direction as the switches of the tap-changing equipment connected to the secondary of the exciting winding are successively closed from position 1 to position 5.

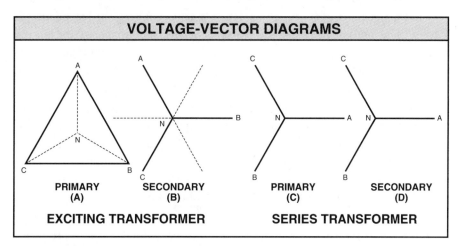

Figure 6-11. A 3φ regulator unit is used to obtain a voltage in the secondary of the series transformer which is at right angles to the normal line-to-neutral voltages.

The primary of the series transformer is also connected in wye. The secondaries of the series transformer windings, which are respectively excited from phases A, B, and C of the exciting transformer, are respectively connected in series with phases B, C, and A of the line in which phase-angle control is desired. This 120° displacement between the primary of the series transformer (AN) and the secondary of the exciting transformer (AN) with the 30° displacement in the opposite direction obtained by the delta-wye connection of the exciting transformer, gives a resultant voltage in the secondary of the series transformer (to be added or subtracted to the line voltage) 90° out-of-phase with the line voltage. Vector AN is 90° to voltage AN.

When switch 1 (position 1) is closed, the voltages of the series transformer are added to those of the line to give a maximum phase displacement of 6° in the positive direction. When switch 3 is closed (position 5), the voltages impressed across the series transformer are zero, therefore there is no displacement of voltage in the line. When switch 5 is closed (position 9), the voltages of the series transformer are added to those of the line to give a maximum phase displacement of 6° in the opposite direction.

A maximum phase control from +6° to −6° (12° total) has been obtained. The actual transitions between these two limits are made in eight equal steps of approximately $1\frac{1}{2}°$ per step.

ϕ is positive (+ϕ) for positions 1 to 5 and negative (−ϕ) for positions 5 to 9. The actual value of the voltage of the secondary of the series transformer for 6° phase displacement is 10.5% of the line voltage. Therefore, a regulating unit for ±6° phase-angle control corresponds very nearly to a regulating unit of 10% buck or boost for voltage control.

By changing the primary connection of the series transformer from wye to delta in the proper sequence with respect to the secondary of the exciting transformer, the voltages in the secondary of the series transformer are in phase with the line-to-neutral voltages, which is the condition required for voltage control. Thus, it is possible to use the same equipment for either voltage control or phase-angle control by proper connection of the series transformer.

Voltage and phase-angle control are often combined in one regulating unit. This requires an additional set of tap-changing equipment, including the preventive autotransformer and another series transformer. The same exciting winding may be used for both circuits. When both voltage and phase-angle control are desired for new transformers, it is more economical to obtain the control directly on the main power transformer.

7 TRANSFORMER CONNECTIONS

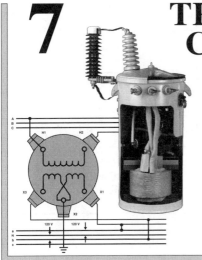

Transformers are used for stepping up or stepping down voltage whether 1φ, 2φ, or 3φ, as well as to obtain various phase transformations. The transformer connections that are possible in power and distribution service for phase transformation include: 3φ to 1φ, 3φ to 2φ or 2φ to 3φ, 3φ to 2φ and 3φ, 2φ to 6φ, and 3φ to 6φ.

TRANSFORMER CONNECTIONS

Almost all electrical power in use today is generated by 3φ generators. In the early development of AC, many 2φ generators were installed and 2φ power was transmitted over 2φ circuits. The secondary circuit of a transformer provided energy for 2φ motors and 1φ lighting and motor loads. Almost all 2φ generators have been replaced with 3φ installations because of the greater efficiency of transmitting power over a 3φ system. A 3φ-to-2φ transformer connection is used to allow the continued use of 2φ motors originally connected to 2φ distribution systems.

VOLTAGE TRANSFORMATION

Although 1φ circuits are not used in the transmission of power nor on the primary side of a distribution circuit, 1φ transformers are used in various applications. The majority of 3φ transmission circuits in the U.S. are made with 1φ transformers connected in various ways to obtain 1φ, 2φ, 3φ, and multi-phase power. Transformers are used for stepping up or stepping down voltage whether 1φ, 2φ, or 3φ, as well as to obtain various phase transformations. The transformer connections that are possible in power and distribution service for voltage transformation include 1ϕ, 2ϕ, and 3ϕ.

Single-Phase Circuits

The low-voltage winding of a standard distribution transformer connected to a 1ϕ circuit is normally made in two equal sections. These sections are arranged so that they may be connected in series or parallel. This arrangement permits current to be delivered at two voltages, one twice the other. See Figure 7-1.

The secondary winding may have its two sections (each of 120 V) connected in series to give a load voltage of 240 V. This connection is used where a motor load is the only load required. The secondary winding may have its two sections connected in parallel to give 120 V. This connection is used primarily for lighting loads, although this voltage may also be used for fractional horsepower motors.

Three transformers may be connected together to furnish power to both motor loads and lighting. However, it is more common to connect the two sections of the low-voltage winding of a transformer in series for 3-wire service. The third wire (neutral) of the 3-wire circuit is connected to the secondary wind-

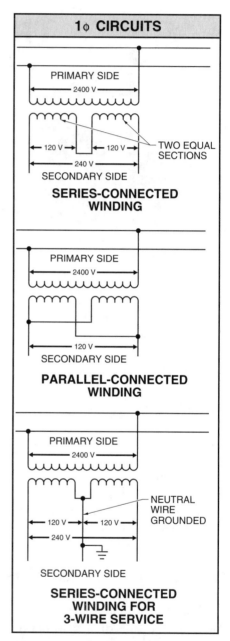

Figure 7-1. The low-voltage winding of a distribution transformer connected to a 1ϕ circuit is normally made in two equal sections connected in series or parallel.

ing at the point where the series connection of the sections is made. This permits the use of the 240 V circuit for a motor load and, at the same time, two 120 V circuits for lighting.

Today, the majority of dwellings are wired with 3-wire service. Electric ranges, dryers, and central climate-control systems are designed for 3-wire 120/240 V operation. In a dwelling, the 3-wire service entrance conductors terminate at the service equipment panel.

Most individual branch circuits carried through dwellings are at 120 V. The circuits which feed large loads, such as air conditioners, electric ranges, or electric water heaters require 240 V, 1ϕ power. The neutral wire carries only the unbalanced current of the entire circuit. No current flows in the neutral wire if each of the 120 V circuits have equal connected loads. If one circuit is fully loaded and the other circuit carries no load, the entire current of the first circuit flows through the neutral wire. If the second circuit is carrying a load less in value than that of the first circuit, the neutral wire carries only the difference in current of the two loads. The 240 V motor load is carried by the two outside wires.

The neutral wire seldom carries as much current as either of the two outside wires. However, the neutral wire should be large enough to safely carry whatever current could flow through the circuit due to a fault, such as a short circuit in either one of the two sides of the 3-wire circuit. Each of the two outside wires should be fused or provided with CBs which operate to protect both the transformer winding and the connected load against undue overloads, short circuits, or grounds.

The neutral wire of a 1ϕ, 3-wire circuit is required to be grounded. Circuits are grounded to limit excessive voltages from lightning, line surges, or unintentional contact with high-voltage lines, and to limit the voltage to ground during normal operation. If the neutral is not grounded and one of the outside wires comes in contact with ground, the other side of the circuit rises to a voltage above ground equal to the series voltage of the two winding sections. A puncture of the insulation between the primary and secondary windings of the transformer causes the secondary winding to have a potential above ground equal to that of the primary winding with which it is in contact. If the neutral is grounded, the low-voltage winding cannot be raised to a potential above ground any higher than the voltage induced in each section.

The four leads of the low-voltage winding sections are brought to three or four bushing leads mounted on the transformer case. Three bushing leads are required if the current-carrying capacity of the bushing lead is sufficient to carry the current of the two sections when connected in parallel. Four leads are brought out of the case when the capacity of the transformer is such that the bushing lead is not large enough to carry the current of the two sections connected in parallel. A transformer is normally connected for 3-wire service by the manufacturer when only three bushing leads are provided. For two-wire service, it is necessary to make the connection on the inside of the case. All connections can be made externally when all four leads are brought out of the case. It is not necessary to remove the cover to see that the transformer is properly connected when all four leads are brought out.

Primary Winding Taps. Taps are normally provided on the primary windings of high-voltage transformers (voltages of 6900 V and above). The main purpose of the taps is to obtain normal secondary voltages when the transformer is used on a line at points where the primary voltage is low due to line drop. These taps may be located at the two ends of the primary winding or in the middle of the winding. See Figure 7-2.

End taps are normally provided on relatively low-voltage windings, with all leads from the windings terminating at a terminal board within the transformer case. Taps located near the center of the winding are provided on the windings in which the voltage ratio is to be changed by means of a tap changer or where the voltage of the winding is relatively high. Additional insulation is required in the winding on

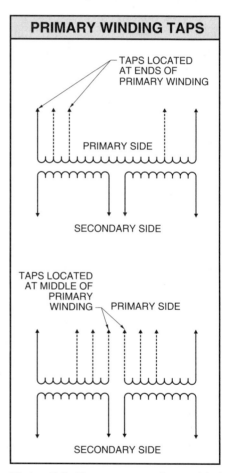

Figure 7-2. Taps are normally provided on the primary windings of high-voltage transformers to obtain normal secondary voltages when the primary voltage is low.

each side of the taps if end taps are used on windings of relatively high voltages (voltages above 13,200 V). The additional insulation is required to protect the winding from line surges.

The voltage ratio of a transformer is changed by connecting the line leads directly to the various tap leads when end taps are used. On transformers provided with midtaps, the line leads are permanently connected to the two ends of the winding. These are properly insulated for voltage surges on the line.

The taps, whether placed near the ends or near the middle of the winding, are often referred to as full-capacity or reduced-capacity taps. Full-capacity indicates that the normal rating of the transformer may be secured when operating on taps. The input current must be increased for the normal rating to be obtained at the tap voltages. This is because of the reduced voltage at which the current is delivered.

With reduced-capacity taps, the normal rating of the transformer decreases by the same percentage by which the tap reduces the voltage of the winding. For this reason, the input current never exceeds the normal current rating. The taps are reduced-voltage taps because they are used on a reduced voltage to maintain constant-rated voltage on the secondary winding. At times, the winding is extended beyond the normal voltage rating of the transformer and taps are placed in this extended winding. *Overvoltage taps* are taps used for voltages above the normal line voltage and maintain a constant-rated voltage on the secondary winding.

Single-Phase Transformer Polarity

Transformer polarity refers to the voltage-vector relationship of the transformer leads as brought outside the transformer tank. Because the high- and low-voltage leads are independent of the arrangement of the windings on the magnetic circuit, the polarity of a transformer can be changed by interchanging the position of the two leads of any one winding as brought out of the tank.

Regardless of which winding is the primary, the voltage-vector relationship is best understood by considering the induced voltages in the high-voltage and low-voltage windings. Because the induced voltages in both coils are induced by the same flux, they must be in the same direction in both windings. Assuming that the induced voltage in the high-voltage winding (H1–H2) is in the same direction as the order of lettering of the leads (H1–H2), the induced voltage in the low-voltage winding flows from X1 to X2.

ANSI standardization rules state that high-voltage leads brought outside the transformer case are to be marked H1, H2, etc., and low-voltage leads are to be marked X1, X2, etc. The order is such that when H1 and X1 are connected and voltage is applied to the transformer, the voltage between the highest numbered H lead and the highest numbered X lead shall be less than the voltage of the full high-voltage winding. When leads are marked in this manner, the polarity of the transformer is subtractive when H1 and X1 are adjacent and additive when H1 is diagonally located with respect to X1. See Figure 7-3.

Assuming a 1:1 ratio of voltage transformation, if the H1 lead is connected to the X1 lead and voltage is applied to the high-voltage winding, a voltmeter connected across leads H2 and X2 reads the difference of the two voltages because they are in opposition. The voltmeter reading is 0 V. For this reason, the polarity of the transformers is subtractive.

If lead H1 is connected to lead X2 and voltage is applied to the high-voltage winding, the voltmeter connected across leads H2 and X1 reads the sum of the two voltages because the two voltages flow in the same direction. The voltmeter reading is twice the voltage applied to the high-voltage winding. For this reason, the polarity of the transformer is additive.

The H1 lead is normally brought out on the right-hand side of the transformer case when facing the high-voltage side to simplify the work of connecting transformers in parallel. The polarity of a transformer is not an indication of the direction of the turns of the high- and low-voltage windings around the core nor does it indicate the voltage stresses that exist between the turns of the two windings.

Subtractive polarity has a small advantage over additive polarity in that the voltage stresses between external leads is smaller. Under normal operating conditions, with leads insulated from each other, the potential stress between adjacent high- and low-voltage leads is one-half the sum of the high and low voltages for additive polarity and one-half their difference for subtractive polarity. The advantage of subtractive polarity, which is ordinarily negligible, becomes appreciable for transformers whose primaries and secondaries have very high voltages.

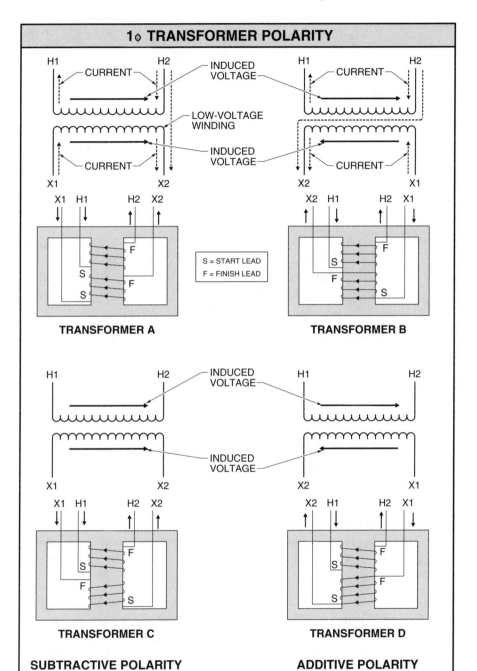

Figure 7-3. Transformer polarity may be subtractive or additive.

The letters S and F indicate the start and finish leads of each of the windings. In transformers that have both windings wound in the same direction with respect to the core of the windings of the two transformers, the polarity is changed from subtractive to additive by interchanging the position of the two X leads.

In transformers that have both windings wound in opposite directions with respect to the core and the same voltage stresses exist between similar parts of the windings of the two transformers, the polarity is changed from additive to subtractive by interchanging the position of the two X leads.

Transformers A and C have subtractive polarity, yet the low-voltage windings are wound in opposite directions with respect to each other. Likewise, Transformers B and D have additive polarity, yet the low-voltage windings are wound in opposite directions with respect to each other. The same polarity in each example is obtained by interchanging the numbering of the start and finish leads of the low-voltage windings.

Transformers having leads marked in accordance with the ANSI rule may be operated in parallel by connecting similarly marked leads provided their ratio, voltages, resistances, reactances, and frequencies permit parallel operation. Also, if the transformers are equipped with tap changers, they must be set on identical taps in each transformer.

The voltages cancel each other and drop to zero if two transformers of opposite polarities but with similar voltage ratios are parallel-connected without regard to the bus connection. In this condition, the circulating current rises to a very high value with chance of burning out both transformers or blowing their fuses.

In general, distribution transformers of 200 kVA or less having voltages of 7500 V or less have additive polarity. All other distribution transformers have subtractive polarity. Some power transformers have additive polarity, although the majority have subtractive polarity.

The terms additive and subtractive are not normally applicable to instrument transformers. The relative instantaneous direction of currents in the leads are required when connecting instrument transformers to wattmeters, watthour meters, power factor meters, etc. The lead polarity of an instrument transformer may also be considered as a description of the relative instantaneous direction of current in its leads. Primary and secondary leads have the same polarity when, at a given instant, the current enters the primary lead and leaves the secondary lead in the same direction, as though the two leads form a continuous circuit. For example, in Transformer A, the broken arrows represent the instantaneous direction of current flow. The current flows in at primary lead H2 and out at secondary lead X2 as though the two leads form a continuous circuit. Similarly, in Transformer B, primary lead H2 and secondary lead X2, although brought out at diagonal corners of the case, are of the same polarity.

To indicate the relative instantaneous direction of current in an instrument transformer, one primary and one secondary lead of the same polarity are marked with a white polarity mark. For example, if Transformers A and B represented instrument transformers, the two leads marked H1 and X1 are marked with a white polarity mark while the other two leads are left unmarked.

Polarity Test Methods

Three methods are commonly used to test for polarity and check transformer lead markings. The methods include transformer comparison, DC use, and AC use. The method used is the most convenient.

Transformer Comparison. The polarity of a transformer may be checked by comparison when a transformer of known polarity and correct lead marking has the same ratio as the one to be tested. The polarity is checked by connecting the high-voltage windings of both transformers in parallel. This is done by connecting the H1 and H2 leads.

If the leads of the transformer to be tested are not marked, assume that similarly located high-voltage leads on the two transformers are of the same marking. Also, connect the left-hand low-voltage leads (facing the low-voltage side) of both transformers, leaving the right-hand side leads free.

Apply a reduced voltage to the high-voltage windings and measure the voltage between the two free leads on the low-voltage side. A reading of 0 V indicates that the relative polarity of both transformers is identical and the lead lettering of the low-voltage winding of the transformer under test is the same as that of the known transformer. A reading equal to the sum of the low voltages of both transformers indicates that the polarities of the two units are opposite and the lead lettering of the low-voltage winding of the transformer under test should be reversed. Before connecting the voltmeter, connect a light fuse or sufficient lamps into the circuit to protect the meter.

DC Use. The polarity of a transformer may be determined by the use of DC. With DC passing through the high-voltage winding, a high-voltage, DC voltmeter is connected across the outlet terminals of the same winding to get a small positive deflection of the voltmeter needle. The two voltmeter leads are transferred directly across the transformer. For example, the lead from the right-hand, high-voltage terminal is placed on the adjacent low-voltage terminal and the other lead is placed on its adjacent low-voltage terminal. The DC excitation is broken, inducing a voltage in the low-voltage winding which causes a deflection in the voltmeter. The polarity is additive if the needle swings in the same direction as previous. The polarity is subtractive if the needle swings in the opposite direction.

AC Use. The polarity of a transformer may be determined by connecting the adjacent left-hand, high-voltage and low-voltage leads (H1 and X1). See Figure 7-4. Apply any convenient value of AC voltage to the high-voltage winding and take readings of the applied voltage and the voltage between the adjacent right-hand, high-voltage and low-voltage leads (H2 and X2). The polarity is subtractive if the voltage between the adjacent right-hand, high-voltage and low-voltage leads is less than the applied voltage. This indicates the difference in voltage between the high-voltage and low-voltage windings. The polarity is additive if the voltage between the adjacent right-hand, high-voltage and low-voltage leads is greater than the applied voltage. This method is limited to transformers in which the ratio of transformation is 30:1 or less.

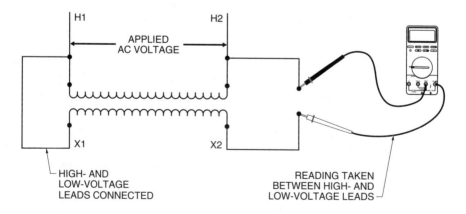

Figure 7-4. The polarity of a transformer may be determined using AC voltage.

Single-Phase Transformer Connections

When both windings of a transformer are wound in the same direction, the directions of the induced voltages must also be in the same direction because the same flux is responsible for both primary and secondary induced voltages. With such an arrangement, adjacent terminals of both windings have similar numbers. Thus, at the instant that the induced voltage is in the direction of H2 to H1 in the primary winding, the secondary voltage is in the direction of X2 to X1. The induced voltage is also in the opposite direction if the secondary winding is wound opposite of the primary. However, this also reverses the subscripts of the secondary winding so that regardless of whether the transformer is of additive or subtractive polarity, the induced voltages are always in the same direction with respect to their terminals.

Additive and subtractive transformers are presented by voltage vectors H1←H2 and X1←X2 because in both cases, the H2 to H1 and X2 to X1 voltages are in phase. The first high-voltage vector may be drawn in any direction. Thereafter, the direction and sense of all other vectors must agree with the direction and sense of the first high-voltage vector. For example, H2 →H1 is another way of drawing H1→H2. This illustrates that the vector has been rotated physically through an angle of 180° without any change of phase in the voltage it represents.

If the vector which represents voltage H2 to H1 is drawn H2→H1, the voltage vector H1 to H2, which is 180° out-of-phase, must be drawn H2←H1. Similarly, if voltage H2 to H1 is drawn H1←H2, the voltage vector for voltage H1 to H2 must be drawn H1→H2. These vectors bear no relationship to the physical arrangement of the terminals on the transformers.

The same terminal (H1 on the left when facing the low-voltage side) is connected to the same line conductor when connecting two 1ϕ transformers to a 1ϕ supply. See Figure 7-5. The low-voltage terminals are adjacent to the high-voltage terminals if the 1ϕ transformers have subtractive polarity.

The X1 terminals are connected together to connect the two 1ϕ transformers in parallel on the low-voltage side. The voltage between the two X2 terminals, when the two X1 terminals are connected together, is determined by adding the two low-voltage vectors. That is, voltage X2 of T1 to X2 of T2 is the vector sum of X2 to X1 of T1 and X2 of T2.

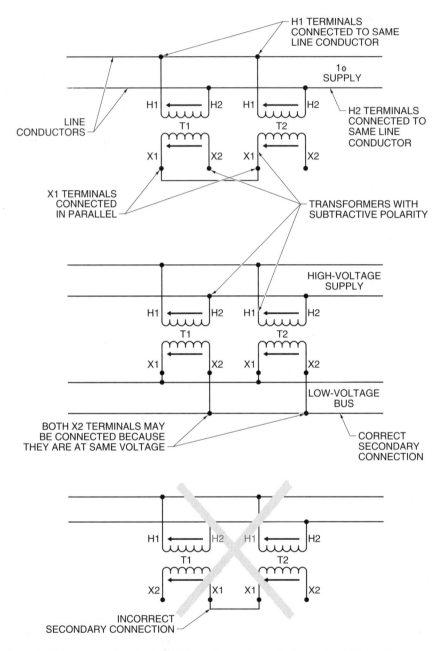

Figure 7-5. To properly parallel 1ϕ transformers, similarly marked high-voltage terminals are connected to the same high-voltage conductor, and similarly marked low-voltage terminals are connected to the same low-voltage conductor.

Assuming that the H2 to H1 vector is H1←H2, then regardless of whether the transformers have additive or subtractive polarity (which determines the physical location of the terminals), the X2 to X1 electrical vectors are X1←X2. A voltage X1 to X2, which is 180° out-of-phase with X2 to X1, may be represented by a vector of the same length but opposite in direction to X2 to X1, namely by X1→X2. Therefore, voltage X1 to X2 of T2 is shown as X1→X2. Voltage X2 to X1 of T1 plus voltage X1 to X2 of T2 equals:

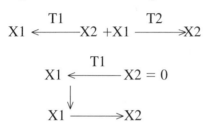

These vectors indicate no voltage difference between terminals X2 of T1 and X2 of T2. Therefore, they may be safely connected.

To properly parallel 1ϕ transformers, similarly marked high-voltage terminals are connected to the same high-voltage conductor, and similarly marked low-voltage terminals are connected to the same low-voltage conductor. A serious condition results if an attempt is made to replace the T1 transformer with one of opposite polarity to T2. The required alterations in the bus connection or retaining the same bus connections must be made.

Two types of connections which may be used when connecting two transformers together for 1ϕ lighting and heating are the double-parallel (banked) connection and the parallel-series secondary connection.

Double-Parallel. The double-parallel (banked) connection is used only for temporary or emergency conditions as a single transformer and is more economical for the same output. See Figure 7-6. The similarly marked terminals are connected to the same line conductors with the addition of the neutral connection. Because the H1 terminals are connected to the A phase, the line conductor, to which the X1 terminals are connected, is also the same phase.

When connecting the secondaries, the voltage between terminals to be connected should be read with a voltmeter before the secondary connections of the second transformer are made to complete the parallel connection. The voltmeter should read zero with the primaries energized because similar terminals are at the same potential if the proposed connection is correct. If there is a voltage reading, the terminals must not be connected and the connection must be changed.

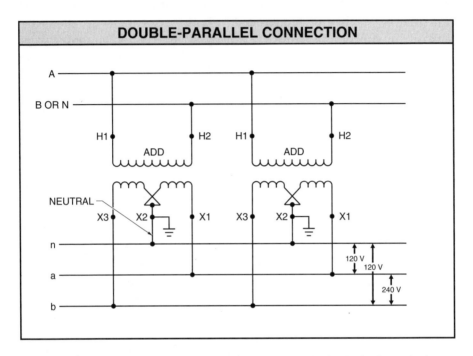

Figure 7-6. In the double-parallel connection, the similarly marked terminals are connected to the same line conductors with the addition of the neutral connection.

There is always a great danger of backfeed when paralleling transformers. Always consider the transformer primary hot, even if the fuse is blown. It is a common industry practice to backfeed a transformer put into service to avoid having an outage. See Figure 7-7.

Transformer B steps the primary voltage down to 240 V and Transformer A acts as a step-up transformer through the low-voltage winding at 240 V. The high-voltage winding of Transformer A is energized at the rated voltage of the high-voltage winding, even though its high-voltage fuses are blown. By backfeed from the secondary of Transformer B, the low-voltage winding of Transformer A becomes the primary winding and Transformer A is energized as a step-up transformer.

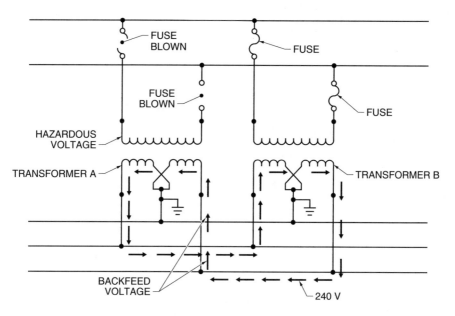

Figure 7-7. Always consider a transformer primary hot because it is a common practice to backfeed a transformer put into service.

Parallel-Series Secondary. The parallel-series secondary connection is used only for temporary or emergency conditions because a single transformer is much more economical. In this connection, each transformer supplies service to only one side of the secondary. When transformers are connected in parallel-series, one transformer does not pick up the load from the other transformer when a fuse is blown. This connection is better when transformers are paralleled without breakers in the secondary between them. See Figure 7-8.

In this circuit, the two adjacent unlike terminals of each transformer are connected together to facilitate a simple connection. This necessitates the connection of the X2 terminal of the left-hand transformer to phase b and the X1 terminal of the right-hand transformer to phase a. The two unlike terminals must be connected together to obtain 240 V across the secondary line conductors.

Little possibility of backfeed exists with this type of connection unless there is a load connected across the hot leads. The disadvantage of using this connection is that when a fuse blows, there is an interruption to a customer's service. This connection can be used when there is not a transformer of sufficient capacity to carry a load, and two smaller transformers are available.

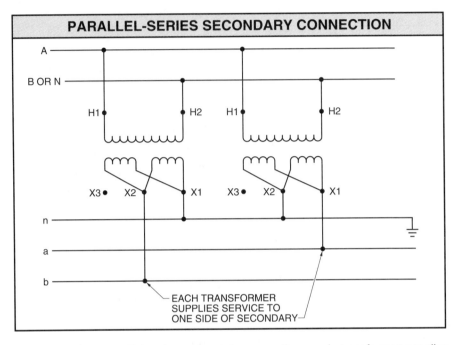

PARALLEL-SERIES SECONDARY CONNECTION

EACH TRANSFORMER
SUPPLIES SERVICE TO
ONE SIDE OF SECONDARY

Figure 7-8. In a parallel-series secondary connection, each transformer supplies service to only one side of the secondary.

Two-Phase Circuits

There are relatively few 2ϕ circuits in existence today. A 2ϕ circuit consists of two 1ϕ circuits that have voltages and currents 90° out-of-phase with each other. See Figure 7-9. Two 1ϕ transformers are used with either the primary, secondary, or both connected for 3-wire or four-wire circuits. The arrangement used depends on whether the two phases are interconnected within the generator or motor windings. For example, if two phases of a generator are connected at their

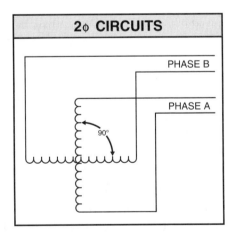

2ϕ CIRCUITS

PHASE B

PHASE A

90°

Figure 7-9. A 2ϕ circuit consists of two 1ϕ circuits that have voltages and currents 90° out-of-phase with each other.

neutral points, a 3-wire circuit on the primary side is not possible. If the phases are entirely free within the generator or motor, any external interconnection required may be made.

The most commonly used 2ϕ connection consists of two 1ϕ transformers connected for 2ϕ, four-wire circuits on both the primary and secondary sides. The voltage-vector diagram of the connection illustrates that phase A and B are 90° out-of-phase with each other. See Figure 7-10.

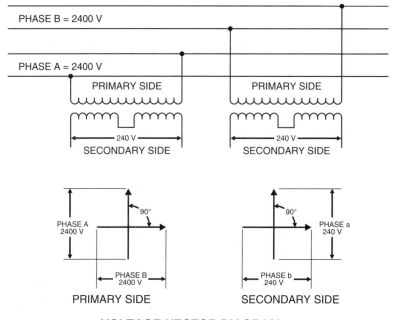

VOLTAGE-VECTOR DIAGRAM

Figure 7-10. The most commonly used 2ϕ connection consists of two 1ϕ transformers connected for 2ϕ, four-wire circuits on both the primary and secondary sides.

Two 1ϕ transformers are connected with one side of the primaries connected to the same line and the other side of the primaries connected to different lines when a 2ϕ, 3-wire circuit is required. See Figure 7-11. In this connection, the voltage across the outside wires is $\sqrt{2}$ (1.41) times the voltage across the winding of each transformer. The middle wire carries the vector sum of the current of the two windings. If the load on each of the two phases is the same, the middle (neutral) wire carries 1.41 times the current of the outside wires. If one of the outside wires of the 3-wire connection is grounded, the voltage to ground becomes 1.41 times the voltage of either phase. If the neutral wire is also grounded, the maximum voltage cannot exceed the phase voltage of either phase.

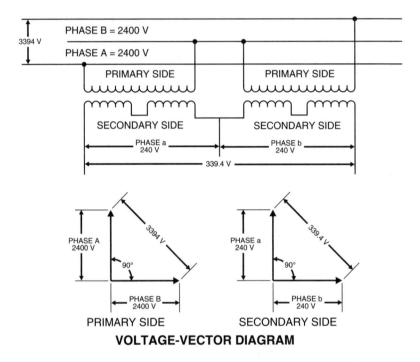

Figure 7-11. Two 1ϕ transformers are connected with one side of the primaries connected to the same line and the other side of the primaries connected to different lines when a 2ϕ, 3-wire circuit is required.

Two 1ϕ transformers may be connected to a 2ϕ supply for obtaining the proper voltage on the secondary side for connection to a rotary converter delivering DC to a 3-wire circuit. In this connection, the midpoints of each secondary of the transformer winding are connected together and grounded, thereby providing the neutral of the 3-wire system which is normally grounded. See Figure 7-12.

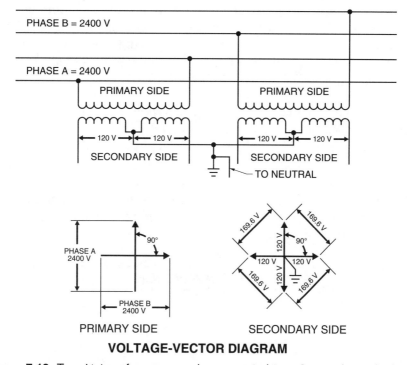

VOLTAGE-VECTOR DIAGRAM

Figure 7-12. Two 1φ transformers may be connected to a 2φ supply to obtain the proper voltage on the secondary side for connection to a rotary converter delivering DC to a 3-wire circuit.

In the previous diagrams, the 1φ transformers have a ratio of 10:1 and a primary voltage of 2400 V. The primary voltage must be equal to the phase voltage of the generator or the lines to which the primaries are to be connected. The ratio depends on the desired secondary load voltage and the primary line voltage.

Three-Phase Circuits

A 3φ system originates from three separate windings of an AC generator or alternator. The three windings are located within the generator to create three separate voltages which are out-of-phase with each other by 120 electrical degrees.

A total of six line leads are required if the three circuits are kept separate from the generator to the apparatus using the current. This is impractical and therefore common wires are used. In forming the common wires, the individual phases of the generator are connected in wye or delta connections.

Wye and Delta Connections. A *wye connection* is a connection in which one end of each of the windings of a polyphase transformer is connected to a common point (the neutral point) and the other end is connected to its appropriate line terminal. A wye (Y) system is also known as the star system. For example, assume that the lines 0A, 0B, and 0C represent the vector relationship of the three voltages generated in the three windings (A, B, and C) of an alternator. The three lines represent the direction and magnitude of the instantaneous voltages of each coil. See Figure 7-13.

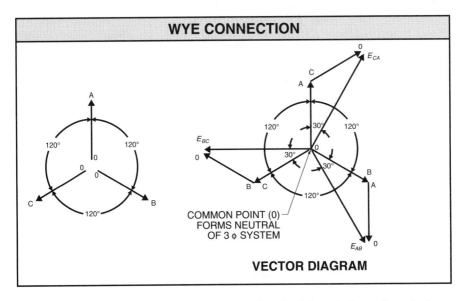

Figure 7-13. A wye connection has one end of each of the windings of a polyphase transformer connected to a common point and the other end connected to its appropriate line terminal.

Assume that the current flowing in each winding, if connected to an external load, is in phase with the voltage generated in each of the respective coils. A 3ϕ wye connection is developed if the three ends (0) are connected and brought out to the remaining three ends (A, B, and C) as line leads. The common point (0) of the three windings forms the neutral of the 3ϕ system. It is the neutral because it is symmetrical with regard to the three coils and may be grounded without creating an unsymmetrical condition. With the three windings connected in wye, the voltage E_{AB} between the two line leads A and B is the vector sum of the coil voltages E_{A0} and E_{0B}. Similarly, the voltage E_{BC} between the two line leads B and C is the vector sum of the coil voltages E_{B0} and E_{0C}. The voltage E_{CA} between the two line leads C and A is the vector sum of the coil voltages E_{C0} and E_{0A}. The vector sum gives the 3ϕ line voltages of the system equal to $\sqrt{3}$ times the voltage of one coil and 120° out-of-phase with each other.

A *delta connection* has the windings of a 3ϕ transformer connected in series to develop a closed circuit. A delta system is developed when the three coils are connected in a three-coil closed circuit and a line lead is connected at each of the three junction points. See Figure 7-14. In this connection, the line voltages are equal to the individual coil voltages. Although the three coils are connected to form a closed circuit, no current flows in the circuit at no load because the vector sum of three equal voltages 120° out-of-phase with each other is always equal to zero. With no voltage impressed across the entire circuit, there can be no current flow. This connection is also known as a mesh system.

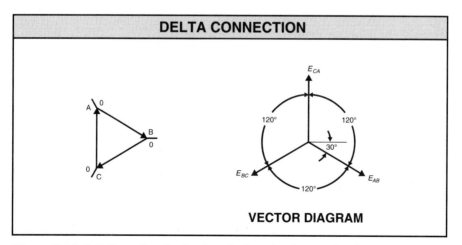

Figure 7-14. A delta system is developed when the three coils of a 3ϕ transformer are connected in a three-coil closed circuit and a line lead is connected at each of the three junction points.

The line current of the 3φ system must be equal to the current in each of the individual coils. However, the line current, as it reaches the junction point of the two windings, has two paths through which it flows. Therefore, the current in the windings of a delta-connected alternator is less than the line current. The winding current is actually 57.7% of the line current (equal to the line current divided by $\sqrt{3}$).

In a wye-connected alternator, the line voltage is equal to the individual coil voltage multiplied by $\sqrt{3}$ and the line current is equal to the coil current. In a delta-connected alternator, the line voltage is equal to the individual coil voltage and the line current is equal to the winding current multiplied by $\sqrt{3}$. The line voltages of a wye-connected alternator are 30° out-of-phase with the line voltages of a delta-connected alternator.

The windings of three 1φ transformers or one 3φ transformer are connected in the same manner as are separate windings of a 3φ generator to make a 3φ group or a single 3φ transformer. The primary and secondary windings may be wye- or delta-connected. The winding combinations include wye-to-wye, wye-to-delta, delta-to-wye, delta-to-interconnected wye, delta-to-delta, open delta, and T-to-T connections. Some of these combinations are approved and in common use. Some combinations have objections which limit their use.

Wye-to-Wye. An advantage of wye-to-wye connections is that the voltage of each 1φ unit connected to form a wye-to-wye connection is only 57.7% of the line voltage. This makes the connection suitable for power transmission because the individual transformers are wound for a relatively low voltage, and nearly double the voltage (173%) is obtained on the line. A wye-to-wye connection also permits the grounding of the neutral on either or both primary and secondary windings. If the primary neutral is carried back to the neutral of the generator, unequal loads may be taken off the secondary windings between the secondary neutral and any of the three line leads. The unbalanced current flows through the neutral back to the generator. This permits lighting loads to be removed at the neutral and any of the 3φ wires and a power load from the 3φ leads to be removed at the same time.

A wye-to-wye connection is used when tying together two high-voltage transmission systems of unequal voltage. In this case, it is necessary that the connection used does not cause a shift in phase from the primary to secondary. The connections that are applicable are the wye-to-wye connection and the delta-to-delta connection. For high voltages, the wye-to-wye connection is preferred because the voltage stresses of the windings to ground with the neutral grounded are only 58% of the voltage stress to ground of similar units connected delta-to-delta. A tertiary winding is provided to suppress the third harmonic voltages which would otherwise appear on the system when a wye-to-wye connection is used for this purpose. A *tertiary winding* is a third winding that is often used in power transformers to provide station power requirements or a tie with synchronous condensers.

A wye-to-wye connection has many disadvantages which restrict its use. In many cases, interference with telephone circuits may result unless a tertiary winding is added to provide a path for the circulating currents to flow through. When using 1ϕ transformers, this trouble may be avoided by grounding the primary and secondary neutrals.

One objection to the use of a wye-to-wye connection is because it is subject to disturbances from harmonic voltages and currents. Another objection is that unbalanced loads cannot be carried on the secondary side unless the primary neutral or fourth wire is provided. Another objection is the fact that it is practically impossible to construct three 1ϕ units or even a 3ϕ unit in which the magnetizing currents of each of the three windings are exactly the same. This makes it impossible to have perfectly balanced voltages in each of the three windings of a wye-to-wye connection.

Three 1ϕ transformers may be connected in a wye-to-wye connection to form a 3ϕ transformer. See Figure 7-15. The 3ϕ voltages on the primary and secondary sides are 4160 V and 416 V respectively if the 1ϕ voltage rating of each transformer is 2400 V to 240 V. The voltage from any line lead to the neutral wire is 2400 V on the primary side and 240 V on the secondary side.

Wye-to-Delta. A wye-to-delta connection, obtained with three 1ϕ transformers or one 3ϕ unit, is used for power transmission and distribution. A wye-to-delta connection permits 1ϕ and 3ϕ loads to be drawn simultaneously from the delta-connected secondary at the same voltage.

A wye-to-delta connection may be made with three 1ϕ transformers. See Figure 7-16. Assuming that only a 1ϕ load is connected across one of the phases, the maximum load that can be obtained without overloading any part of the circuit is 150% of the maximum rating of the 1ϕ transformer.

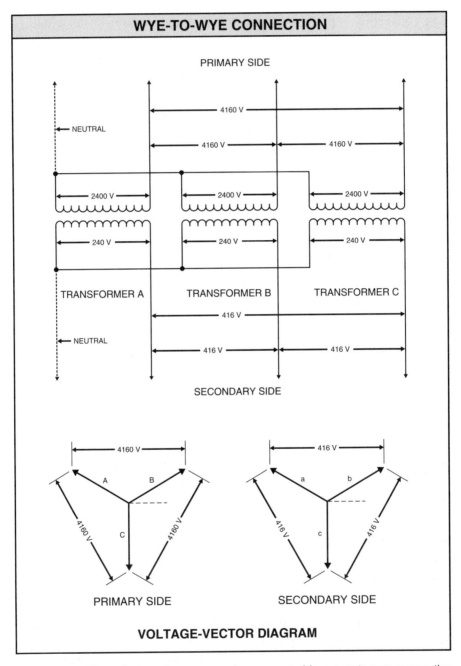

Figure 7-15. Three 1ϕ transformers may be connected in a wye-to-wye connection to form a 3ϕ transformer.

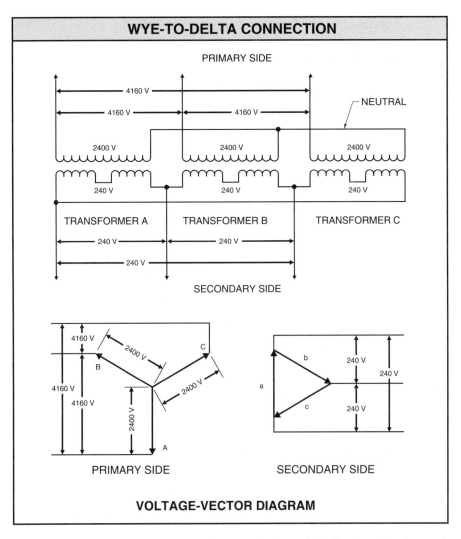

Figure 7-16. A wye-to-delta connection permits 1φ and 3φ loads to be drawn simultaneously from the delta-connected secondary at the same voltage.

With simultaneous 1φ and 3φ loading, the currents of the two loads are added in their proper phase relations to determine the maximum loads that can be drawn from the connection. The resultant current in any one of the three transformers due to this loading must not continuously exceed the maximum current rating of the winding.

In a wye-to-delta connection of three 1ϕ transformers, one unit may be disconnected from the circuit and service maintained with the secondary operating in open delta at 57.7% of normal bank capacity. This is provided the neutral on the primary side and the neutral of the supply source for the primary are grounded. See Figure 7-17. The system is unbalanced and considerable telephone interference may result from such a connection.

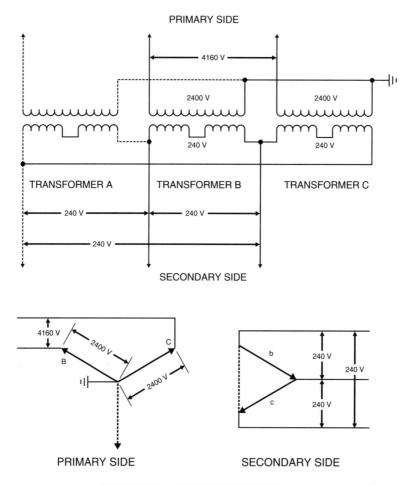

VOLTAGE-VECTOR DIAGRAM

Figure 7-17. In a wye-to-delta connection of three 1ϕ transformers, one unit may be disconnected from the circuit and service maintained with the secondary operating in open delta at 57.7% of normal bank capacity.

Delta-to-Wye. A 3φ delta-to-wye connection is often used for distribution where a four-wire secondary distribution circuit is required. The delta-connected primary permits 1φ loads to be connected between the secondary neutral and the 3φ line leads. The 3φ wye voltage is $\sqrt{3}$ times the 1φ (line-to-neutral) voltage. See Figure 7-18.

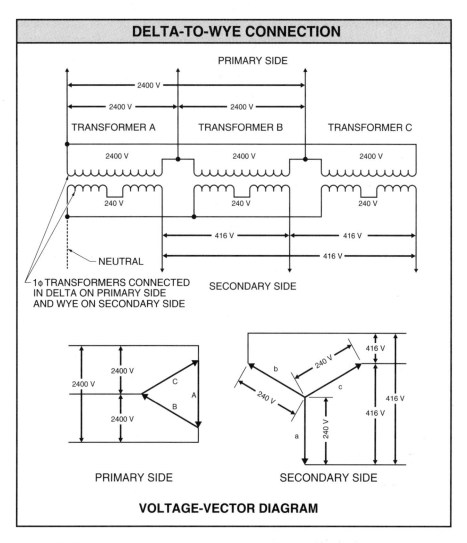

Figure 7-18. A 3φ delta-to-wye connection is often used for distribution where a four-wire secondary distribution circuit is required.

A delta-to-wye connection, with the neutral of the wye connection available, is sometimes used for supplying power to a synchronous converter delivering DC for a 3-wire circuit. The third wire of the 3-wire DC circuit is obtained by connecting the third wire of the system directly to the neutral of the wye-connected secondary of a delta-to-wye bank of transformers. See Figure 7-19. The unbalanced direct current (*3I*) divides at the neutral point of the wye connection and one-third flows through each of the transformer secondary windings. This increases the magnetic density of the transformer and increases the exciting current and iron loss. Therefore, this connection should be used only on relatively small capacities of rotary converters where the maximum unbalanced DC flowing through each transformer is not in excess of 10% of the 1ϕ current rating.

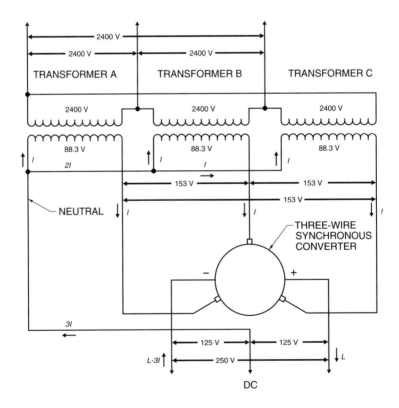

Figure 7-19. A delta-to-wye connection is sometimes used for supplying power to a synchronous converter delivering DC for a 3-wire circuit.

In a 3ɸ core-type transformer connected in wye on its secondary side for synchronous converter service, the magnetizing action of the unbalanced DC is practically negligible. This is because the DC flows in the same direction in each of the three legs from one end of the core to the other end. The fluxes established by the current must use the surrounding media, such as air or oil and transformer tank for their return path. The reluctance of such a path is very high. Therefore, the magnetizing effect of the DC is small.

Delta-to-Interconnected Wye. Synchronous converters are frequently connected to a bank of transformers connected delta-to-interconnected wye to eliminate the flux distortion in the transformer due to the unbalanced DC in the neutral of the 3-wire circuit flowing through the windings. Two separate interconnected windings are used on each phase of the wye connection and the windings are connected so that the DC flows in opposite directions around the two halves of each phase winding, neutralizing the flux distortion. The interconnected wye connection is often referred to as a zigzag winding.

Three 1ɸ transformers (A, B, and C) may be connected in delta on the primary side and in interconnected wye on the secondary side. See Figure 7-20. The secondary winding of each transformer is composed of two equal windings and the two halves of each winding are connected in different legs of the interconnected wye. Each leg of the interconnected wye comprises two half-sections connected in opposition. The half-sections are 120° out-of-phase with each other.

Each leg voltage is the vector difference of the two half-section voltages and is numerically equal to $\sqrt{3}$ times the voltage of one half-section. The voltage across the terminals of the interconnected-wye connection is the vector difference of two leg voltages 120° out-of-phase with each other. This difference equals $\sqrt{3}$ times the voltage of each leg. Because each leg voltage is $\sqrt{3}$ times the voltage of each half-section, the terminal voltage of the interconnected-wye connection must be three times the voltage of each half-section winding. This is shown in the voltage vector diagram in which the primary side of the transformer is connected in delta to a 2400 V supply, and the secondary is connected in interconnected wye to a 250/125 V, 3-wire synchronous converter. Assuming that the DC in the positive line of the 250 VDC circuit is *L* amps and the current in the negative line is *L-3I* amps, there is an unbalanced current of *3I* amps in the neutral.

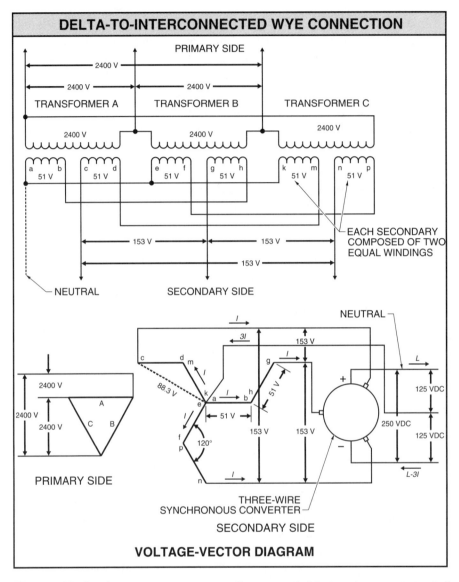

Figure 7-20. Synchronous converters are often connected to transformers connected delta-to-interconnected wye to eliminate the flux distortion in the transformer due to the unbalanced DC in the neutral of the 3-wire circuit flowing through the windings.

This unbalanced current flows into the neutral of the interconnected wye windings and divides equally through each of the three legs. The direction of the current flow in the half-section a-b is opposite in direction to the flow of the current in the half-section c-d. The magnetizing action of the unbalanced current is neutralized because the two half-sections a-b and c-d form the two half-sections of the secondary winding of Transformer A.

The low-voltage side operates at only 86.6% of its normal capacity (if operated in straight wye connection) because the two half-sections of each transformer secondary winding are connected in different phases. Therefore, transformers arranged for interconnected wye operation are larger than those arranged for straight wye connection. The actual ratio of kilovolt-amps of transformer capacity to kilovolt-amps transformed for the interconnected wye system is 1.075:1.

Delta-to-Delta. Three-phase transformers are seldom connected in a delta-to-delta connection. However, 1ϕ transformers connected to form a 3ϕ bank are quite common.

Delta-to-delta connected transformers must be wound for full-line voltage. A 3ϕ delta-to-delta connection and voltage-vector diagram show that the voltage of each transformer is equal to the 3ϕ line voltage. The current in each of the transformers is only 57.7% of the line current. The windings have a greater number of turns than for a wye connection of the same line voltage, while the cross section of the turns is only 57.7% of that for a wye-connected transformer. See Figure 7-21.

The midpoint of one of the windings may be grounded for 240/120 V, 1ϕ, 3-wire service for lighting if the secondary is wound in two sections. See Figure 7-22. Occasionally, the 1ϕ unit that is to be used for the lighting load is made larger than the other two units so that the maximum kilovolt-amps available in the other two units for 3ϕ loading may be used. If the midpoints are available on the other two phases, 240/120 V, 1ϕ, 3-wire service may also be taken from these circuits. The middle wire of each of the two phases must not be grounded because this short circuits the phases. A voltage equal to one-half of the line voltage existing between the midpoints and 3ϕ loads is obtained. Simultaneous loading at full voltage and half-voltage is possible. The maximum load that can be obtained at half-voltage, with no load at full voltage, is equal to one-half the bank capacity of the three transformers.

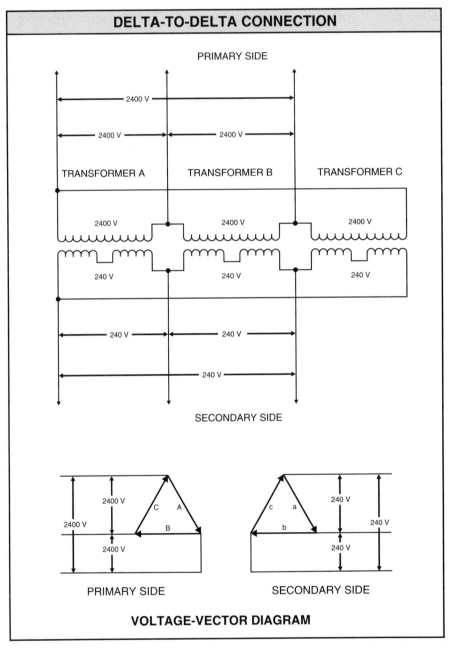

DELTA-TO-DELTA CONNECTION

PRIMARY SIDE

2400 V

2400 V — 2400 V

TRANSFORMER A — TRANSFORMER B — TRANSFORMER C

2400 V — 2400 V — 2400 V

240 V — 240 V — 240 V

240 V — 240 V

240 V

SECONDARY SIDE

2400 V — C — A
2400 V — B
2400 V

PRIMARY SIDE

240 V — c — a
240 V — b
240 V

SECONDARY SIDE

VOLTAGE-VECTOR DIAGRAM

Figure 7-21. In a delta-to-delta connection, the voltage of each transformer is equal to the 3φ line voltage and the current in each of the transformers is only 57.7% of the line current.

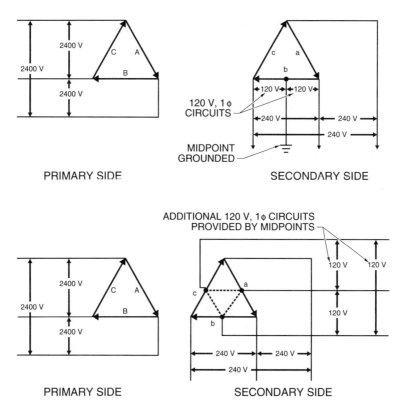

Figure 7-22. The midpoints of the windings provide 240/120 V, 1φ, 3-wire service for lighting applications.

Transformers operating in a 3φ delta-to-delta connection have a circulating current flowing in both the primary and secondary windings if their ratios are different. This fails to divide the load properly in each of the windings if their respective impedances and the ratios of reactances to resistances are not equal.

Open-Delta. An open-delta connection makes it possible to maintain operation if one 1φ transformer or one winding of a 3φ shell-type transformer in a delta-to-delta system becomes defective. See Figure 7-23. The defective unit is removed from the circuit on the primary and secondary sides. In the case of the 3φ transformer, the defective phase should be disconnected from the circuit and the primary and secondary windings of the defective phase should be short circuited to prevent stray fluxes from the other phases inducing voltage in the faulty windings. The capacity of the two units or two windings of a 3ϕ unit connected in open-delta connection is 57.7% of the normal 3ϕ bank capacity.

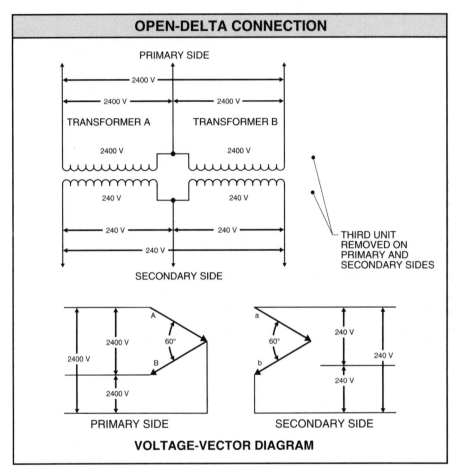

Figure 7-23. An open-delta connection enables continued operation if one 1ϕ transformer or one winding of a 3ϕ shell-type transformer in a delta-to-delta system becomes defective.

An open-delta connection can be used to supply a 3ϕ load which is currently light but is expected to increase in the future. This installation requires two 1ϕ transformers which keeps the initial investment low but also provides for future increase of load.

For example, a transformer bank is required for a new industrial building that has an initial 3φ load of 150 kVA with an anticipated future load of 300 kVA. If two 1φ, 100 kVA transformers are installed and connected in an open-delta connection, the present 3φ capacity of the transformer bank is 57.7% of the normal 3φ bank capacity or 57.7% of 300 kVA which provides a present capacity of 173 kVA. As the load increases to the 173 kVA open-delta capacity, a third 100 kVA transformer may be installed to increase the capacity of the delta-to-delta connected bank to 300 kVA.

An open-delta connection is also used when the secondary circuits are to supply lighting and power loads. The grounded or neutral conductor of the lighting circuit is taken from a center tap of the 240 V secondary winding. This provides a 120/240 V, 1φ, 3-wire lighting circuit. When an open-delta connection supplies a large 1φ load and a small 3φ load, the two transformers are of different kVA ratings. The 1φ load transformer is larger than the 3φ load transformer. This application is economical because one small 1φ transformer can be added to a large 1φ transformer to supply a limited 3φ power load.

T-to-T. A T-to-T connection requires two 1φ transformers. One of the units is the main transformer and is provided with a 50% voltage tap to which a teaser transformer is connected. A *teaser transformer* is a transformer connected between the midpoint of the main transformer and the third-phase wire of the 3φ system. The teaser transformer may be designed for 86.6% of the line or main transformer voltage, but normally is made identical with the main transformer and operated at reduced flux density. Thus, it is possible to operate two identical 1φ transformers connected in a T-to-T connection or open-delta connection if one of the transformers in the delta-to-delta connected bank becomes inoperative. The only requirement for the T-to-T connection is the 50% tap. See Figure 7-24.

The bank capacity of a T-to-T connection is the same as for two transformers connected in an open-delta connection. The efficiency is somewhat higher because the teaser transformer, if identical units are used, operates at reduced flux density and has lower iron loss. The increase in efficiency, due to lower iron loss, should be considered when a bank of transformers is to be operated for several years.

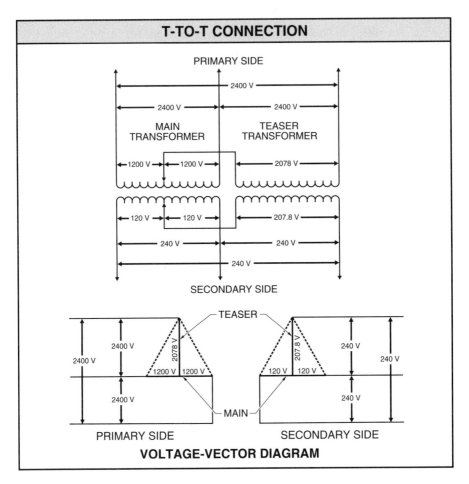

Figure 7-24. In a T-to-T connection, one of the units is the main transformer and is provided with a 50% voltage tap to which a teaser transformer is connected.

A T-to-T connection may be used for operating a synchronous converter, provided a neutral is brought out of the teaser winding at a point equal to one-third of the total winding measured from the end connected to the main transformer. The magnetizing action of the unbalanced DC flowing through the transformer windings is neutralized in each of the two transformers and there is no increase in iron loss due to the DC flow. See Figure 7-25.

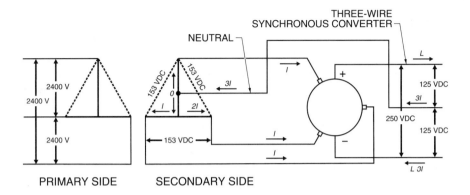

Figure 7-25. A T-to-T connection can be used for operating a synchronous converter, provided a neutral is brought out of the teaser winding at a point equal to one-third of the total winding measured from the end connected to the main transformer.

In this circuit, *3I* represents the unbalanced current in the neutral of the 3-wire DC system. At point 0, the current divides, two-thirds (*2I* amps) flowing toward the main transformer and one-third (*I* amps) flowing out of the line end of the teaser transformer and returning to the converter circuit. At the junction of the main and teaser windings, the *2I* amps divide and *I* amps flow through each of the two half-sections of the main transformer. In the main transformer, the DC (*I* amps) is flowing in opposite directions. For this reason, the fluxes established by them neutralize each other. In the teaser winding, *2I* amps flow through one-half as many turns and in the opposite direction to that of the *I* amps. Therefore, the magnetomotive forces of the DC flowing in the teaser winding also neutralize each other.

PHASE TRANSFORMATION

Transformers are used for stepping up or stepping down voltage, whether 1ϕ, 2ϕ, or 3ϕ, as well as to obtain various phase transformations. The transformer connections that are possible in power and distribution service for phase transformation include: 3ϕ to 1ϕ, 3ϕ to 2ϕ or 2ϕ to 3ϕ, 3ϕ to 2ϕ and 3ϕ, 2ϕ to 6ϕ, and 3ϕ to 6ϕ.

Three-Phase to Single-Phase

It is nearly impossible to deliver 1ϕ current from a 3ϕ source of supply by means of transformer action alone and have balanced conditions. This is true because 1ϕ power is pulsating (it passes from a maximum value through zero and back to a maximum every half-cycle). Three-phase power is continuous. It is delivered at a constant rate. For these reasons, it would be necessary for the system to store energy during the interval of time when the power delivered to the 1ϕ side is less than the power received from the 3ϕ side. Transformers have no capacity for storing energy at a given time and subsequently transmitting it. For this reason, 3ϕ to 1ϕ transformation is impossible.

Various schemes have been proposed for the transformation of 3ϕ to 1ϕ, but none gives better results than connecting a 1ϕ transformer across one of the three phases. A wye-connected source of supply allows an equal current in two of the phases and zero current in the third. With a delta-connected source, two of the phases have the same current and the third has a current twice as large as that in the other two.

Three-Phase to Two-Phase

Three-phase to 2ϕ or 2ϕ to 3ϕ transformation may be accomplished in several different ways. The most common connection is the Scott connection.

Scott Connection. A Scott connection requires two transformers, one being the main and the other the teaser. The windings of each transformer are the same on the 2ϕ side. On the 3ϕ side, the number of turns in the teaser winding is 86.6% of the number of turns in the main winding. On the 3ϕ side, a 50% tap is brought out of the main winding. Both transformers are made identical with respect to the 3ϕ side for interchangeability and replacement if one of the transformers becomes inoperative. The 3ϕ windings are wound with the same number of turns and 50% and 86.6% taps are placed in each of the windings. When operating as a main winding, the 86.6% tap is not in use. When operating as a teaser, the 50% tap and the portion of the winding beyond the 86.6% tap are not in use. See Figure 7-26.

The voltage-vector diagram shows the voltage conditions of a 3ϕ to 2ϕ transformation in addition to a voltage transformation from 2400 V to 240 V. The two half-sections of the 3ϕ winding on either side of the 50% tap should be distributed over the entire winding length of the magnetic circuit to prevent flux distortion and poor regulation.

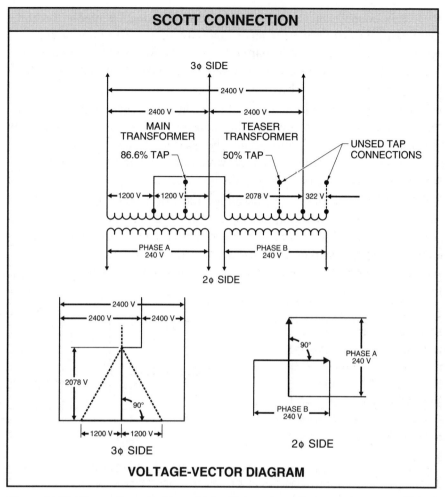

Figure 7-26. Three-phase to 2φ or 2φ to 3φ transformation may be accomplished using the Scott connection.

In emergency cases where a transformer with an 86.6% tap is not available, a T-to-T connection, in which the two transformers are identical with respect to their voltage ratio, may be used. This is an unsymmetrical connection and, when used, no attempt should be made to use it in parallel with a true 3φ circuit or balanced 2φ circuit.

If this connection is resorted to in transforming 2ϕ to 3ϕ, the three phases are not exactly 120° apart, and two of the 3ϕ voltages are 12% greater than the third. When transforming 3ϕ to 2ϕ with a T-to-T connection, the 2ϕ voltages are in quadrature, but the one phase has a voltage 15% greater than the other.

This unsymmetrical condition, due to an emergency connection, may be lessened if a transformer having reduced-voltage taps is available. In such a case, the transformer should be used as the teaser and a 3ϕ connection should be made on that tap which most nearly approaches 86.6% of the total winding on the 3ϕ side.

Two 1ϕ transformers with a voltage ratio of 1:10 (240 V to 2400 V) may be connected in a T-to-T connection for emergency transformation from 2ϕ to 3ϕ. See Figure 7-27. The T-to-T connection is exactly the same as the Scott connection except that the teaser transformer is connected across 100% of the main winding instead of on the 86.6% tap.

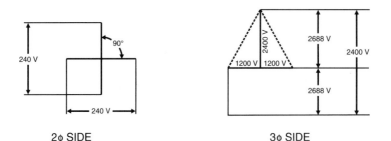

Figure 7-27. Two 1ϕ transformers with a voltage ratio of 1:10 (240 V to 2400 V) may be connected in a T-to-T connection for emergency transformation from 2ϕ to 3ϕ.

The 2ϕ windings are electrically independent in the Scott connection. There are several connections in which the windings are electrically interconnected on the 2ϕ side.

A 3ϕ to 2ϕ transformation may be accomplished using three 1ϕ transformers with the 2ϕ side interconnected. See Figure 7-28. The 3ϕ side is connected in delta. With the proper voltage transformation, the 3ϕ side may be connected in wye. This connection requires a 50% tap in one winding and an 86.6% tap on each of the other two windings on the 2ϕ side.

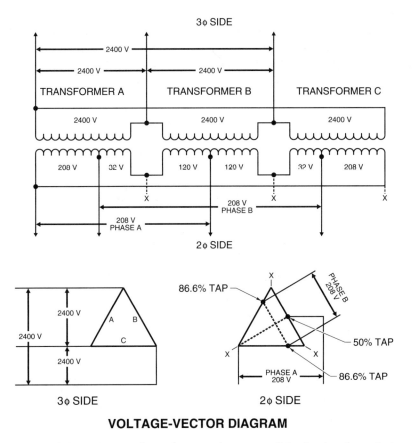

VOLTAGE-VECTOR DIAGRAM

Figure 7-28. A 3φ to 2φ transformation may be accomplished using three 1φ transformers with the 2φ side interconnected.

Instead of using three 1φ transformers, this transformation may be obtained with a 3φ magnetic circuit. In this case, each transformer represents the primary and secondary windings of one of the three legs of the 3φ transformer.

Three-Phase to Two-Phase and Three-Phase

Three-phase power may be drawn from transformers simultaneously with 2φ power by connections made to the corners of the delta connection on the 2φ side. In this case, the voltage of the 3φ circuit is higher than the voltage of the 2φ circuit. If the 2φ voltage is 208 V, the 3φ voltage is 240 V. To obtain the same 2φ and 3φ voltages on the secondary side, two 1φ transformers connected in a T-to-T and Scott connection are used. See Figure 7-29. The T-to-T connection is used for the 3φ transformation and the Scott connection for the 3φ to 2φ transformation.

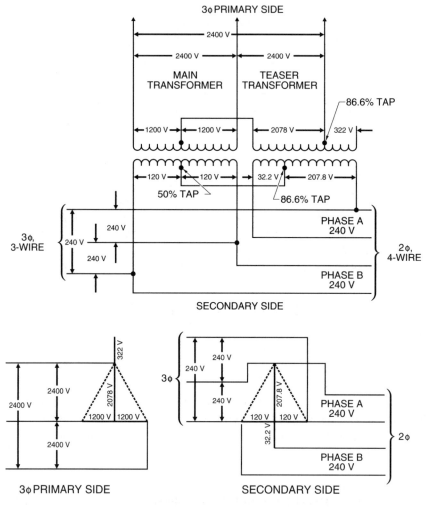

VOLTAGE-VECTOR DIAGRAM

Figure 7-29. Three-phase power may be drawn from transformers simultaneously with 2φ power by connections made to the corners of the delta connection on the 2φ side.

Two-Phase to Six-Phase

Two 1φ transformers connected in a double Scott connection are used when operating a 6φ synchronous converter from a 2φ circuit. This connection requires special units of the same impedance. Each must have two low-voltage windings connected so that their voltages are 180° from each other in phase to produce the 6φ system. Normally, the synchronous converter is started from the AC side as a 3φ converter. This further complicates the transformer windings.

Additional complications are introduced if a neutral must be provided. When all of these conditions must be met in the design of the transformers, they can no longer be made identical as in the case of the single Scott connection. See Figure 7-30. The 2φ primary windings may be connected for either 3-wire or 4-wire service. The numbers 1 to 6 represent the six successive collector rings of the synchronous converter.

Three-Phase to Six-Phase

The four common connections for transforming from 3φ to 6φ for synchronous converter service are the T-, diametrical, double-delta, and double-wye connections.

T-Connections. A T-connection is made with two 1φ transformers having their primaries connected in a T-to-T connection and the secondary windings connected so that their voltages are 180° from each other in phase. In this connection, a 2400 V, 3φ primary is assumed and the voltages of the windings on the secondary side are assumed to give the proper voltage across the six rings of a 250 V or 250/125 V, 6φ, 3-wire synchronous converter.

Diametrical Connections. A diametrical connection is the most commonly used of all 3φ to 6φ transformations. It requires only one low-voltage winding on each 1φ transformer. The two secondary leads are connected to diametrically opposite points on the converter's armature windings. The middle points of the three low-voltage windings may be connected to form a neutral for the 3-wire circuit. This neutral connection should be arranged so that it can be opened during the time that the synchronous converter is started from the AC side.

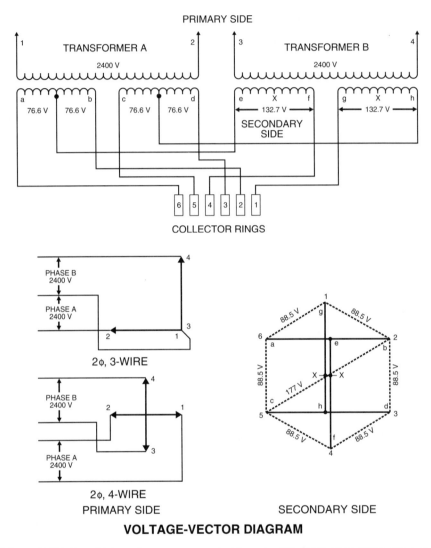

Figure 7-30. Two 1ɸ transformers connected in a double Scott connection are used when operating a 6ɸ synchronous converter from a 2ɸ circuit.

Should one of the transformers become inoperative, it is possible to continue operation at reduced capacity with the other two units connected across their respective converter diameters. The high-voltage windings may be connected in wye or delta, although a delta-connected primary is preferred because of the third harmonic voltage introduced in a wye-to-wye connection. See Figure 7-31.

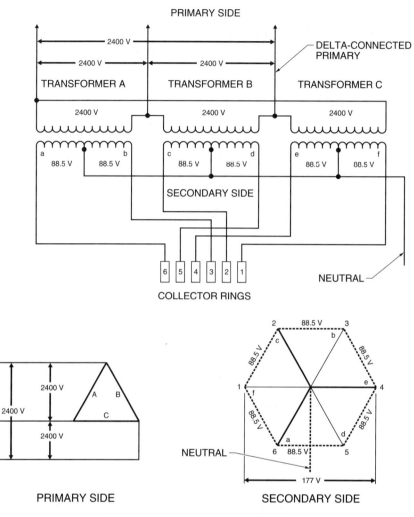

VOLTAGE-VECTOR DIAGRAM

Figure 7-31. A diametrical connection requires only one low-voltage winding on each 1ϕ transformer.

In a 6ϕ diametrical connection with common neutral, one-half the output can be taken from the low-voltage side for operating 3ϕ devices without a change of diametrical voltage. A 3ϕ, 250/125 V converter may be operated from one-half of the winding.

The low-voltage coils are connected in a delta-connection if full 3ϕ output is required. In this case, the diametrical voltage is increased 15.4%. See Figure 7-32. In this circuit, the DC voltage is equal to 288.5 V. The DC voltage is equal to the diametrical voltage multiplied by $\sqrt{2}$ regardless of the number of phases.

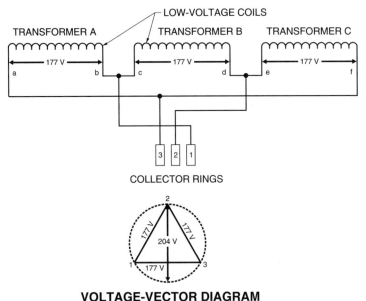

VOLTAGE-VECTOR DIAGRAM

Figure 7-32. The diametrical voltage is increased 15.4% by connecting the low-voltage coils in a delta-connection for a 3ϕ output.

In a 6ϕ converter, the diametrical voltage is equal to the voltage across rings 1-4, 2-5, and 3-6. These rings are connected to diametrically opposite points of the converter winding.

The full 3ϕ output may also be obtained by connecting the low-voltage windings in a wye connection. In this case, the diametrical voltage is $\sqrt{3}$ times the voltage of the delta-connected windings. The neutral of the wye connection should be grounded when connected for the increased voltage.

Double-Delta Connections. A double-delta connection is normally used when full output is required at the same voltage at either 3ϕ or 6ϕ. A double-delta connection requires two separate low-voltage windings on each transformer.

Both sets of windings are connected in a delta connection, but one is reversed with respect to the other so that two delta connections, displaced 180° from each other, are formed. The high-voltage windings may be connected in delta or wye connections. The delta connection is preferred because it permits continued operation at reduced capacity with one of the units disconnected from the circuit if it becomes inoperative. See Figure 7-33.

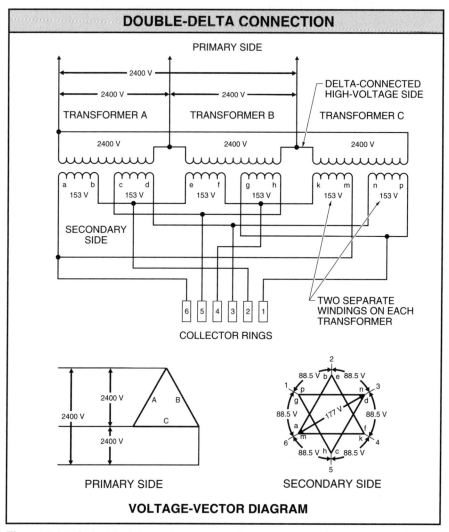

Figure 7-33. A double-delta connection requires two separate low-voltage windings on each transformer.

By connecting the two low-voltage windings of each transformer in parallel, full output at 3ϕ is obtained when the parallel groups are connected in a delta connection. See Figure 7-34. The diametrical voltage is unchanged by this connection. For this reason, a double-delta connection is commonly used for 3ϕ or 6ϕ transformation at the same voltage.

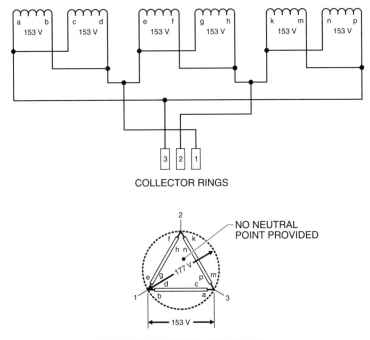

COLLECTOR RINGS

VOLTAGE-VECTOR DIAGRAM

Figure 7-34. Connecting the two low-voltage windings of each transformer in parallel produces full output at 3ϕ when the parallel groups are connected in a delta connection.

A double-delta connection cannot be used for 3-wire service because no neutral point is available. Separate autotransformers must be connected across the delta-connected windings if a neutral point is required.

Double-Wye Connections. A double-wye connection, like a double-delta connection, requires two sets of low-voltage windings displaced 180° in phase with each other. See Figure 7-35. Likewise, the high-voltage windings may be connected either in delta or wye, although the delta connection is preferable.

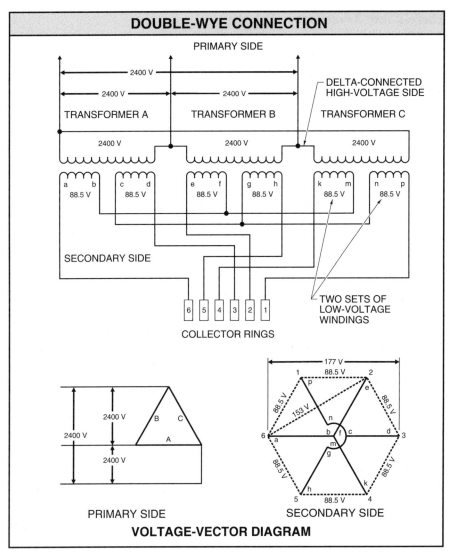

DOUBLE-WYE CONNECTION

PRIMARY SIDE

2400 V

2400 V — 2400 V

DELTA-CONNECTED
HIGH-VOLTAGE SIDE

TRANSFORMER A   TRANSFORMER B   TRANSFORMER C

2400 V   2400 V   2400 V

a   b   c   d   e   f   g   h   k   m   n   p
88.5 V   88.5 V   88.5 V   88.5 V   88.5 V   88.5 V

SECONDARY SIDE

6 5 4 3 2 1

TWO SETS OF
LOW-VOLTAGE
WINDINGS

COLLECTOR RINGS

2400 V   B   C
2400 V   A
2400 V

177 V
1   88.5 V   2
88.5 V   153 V   n   88.5 V
6   b f c   3
a   m
g
88.5 V   88.5 V
h   k
5   88.5 V   4

PRIMARY SIDE   SECONDARY SIDE

VOLTAGE-VECTOR DIAGRAM

Figure 7-35. A double-wye connection requires two sets of low-voltage windings displaced 180° in phase with each other.

ANSI standards give recommendations and standard practices for 1ϕ, 3ϕ, and 6ϕ connections showing polarity, relative phase rotation, angular displacement, and proper lead markings. See Appendix.

Three-Phase Transformer Polarity

In polyphase transformers, polarity alone cannot represent a definite relation between the high- and low-voltage windings. In addition to the lead markings, voltage-vector diagrams are required to show the angular-phase displacement between the high- and low-voltage windings and the time order of phase sequence.

As a rule, all phases of a 3ɸ transformer have the same relative polarity (expressed in terms of 1ɸ polarity). If the polarity of one phase is subtractive, the polarity of the other two phases is subtractive. If the polarity of one phase is additive, the polarity of the other two phases is additive. The 3ɸ polarity is dependent on the interphase connections of the respective outlet leads to the full-phase windings as well as on the polarity of the separate phases.

The three high-voltage leads and the three low-voltage leads which connect to the full-phase windings are marked H1, H2, H3, and X1, X2, and X3 respectively. The markings are applied so that with the phase sequence of voltage on the high-voltage side in the time order H1, H2, H3, on the low-voltage side, it is in the time order of X1, X2, X3. In the low-voltage vector diagram of three 1ɸ transformers connected in a delta-to-delta 3ɸ bank, the phase rotation or phase sequence is in a clockwise direction for both the high-voltage and low-voltage sides. See Figure 7-36. The phase rotation is only relative. The actual phase rotation is dependent on and equal to the phase rotation of the supply voltage. If a 3ɸ motor were first connected across the three leads of the high-voltage side and then transferred directly to similarly numbered leads of the low-voltage side, that is, the motors leads are transferred from H1 to X1, H2 to X2, and H3 to X3, the motor would continue to rotate in the same direction.

The angular-phase displacement between the high- and low-voltage windings as defined by ANSI is the angle between the lines H1-N and X1-N where N is the neutral point of the voltage-vector diagram. In this circuit, the angle for the connection is 0°. Although three 1ɸ transformers are shown, the three transformers may be considered as the windings of each of the three legs of a 3ɸ transformer because the connections are identical.

The angular-phase displacement of the 3ɸ transformer (or three 1ɸ transformers) varies based on the four combinations in which the transformer(s) may be connected. The angular-phase displacement depends on the combination used.

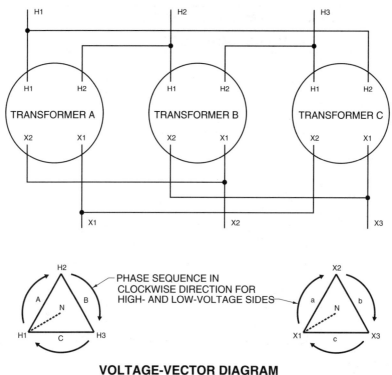

VOLTAGE-VECTOR DIAGRAM

Figure 7-36. In the voltage-vector diagram of three 1φ transformers connected in a delta-to-delta 3φ bank, the phase rotation is in a clockwise direction for the high- and low-voltage sides.

Transformers connected in a delta-to-delta or wye-to-wye connection may have an angular-phase displacement of 0° or 180°. Three 1φ transformers of additive polarity may be connected in a delta-to-delta connection with 180° angular-phase displacement. The angular-phase displacement of 180° is obtained by reversing the delta connection on the low-voltage side.

Wye-to-delta-connected transformers, with properly marked leads, always have an angular-phase displacement of 30°. For this reason, all 3φ to 3φ or 3φ to 6φ connections can be grouped in one of five different groups having an angular-phase displacement of 0°, 30°, or 180°. See Figure 7-37. The vector diagram of 3φ transformers for 3φ to 3φ or 3φ to 6φ operation is normally included by the manufacturer in the markings on the nameplate or diagram of connections which forms part of the transformer.

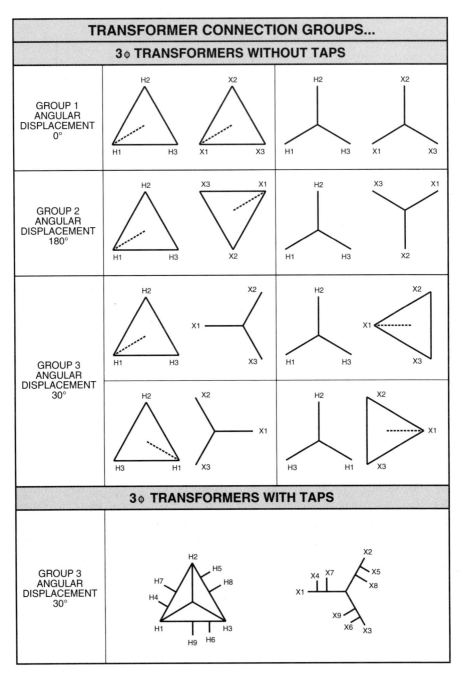

Figure 7-37 continued...

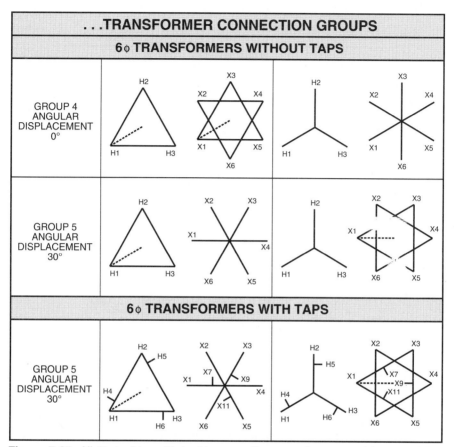

Figure 7-37. All 3φ to 3φ or 3φ to 6φ connections can be grouped in one of five different groups, having an angular displacement of 0°, 30°, or 180°.

Three-Phase Transformer Voltage-Vector Diagram Construction

The polarity of each phase of a 3φ transformer may be determined in the same manner as 1φ transformers. The voltage-vector diagram is constructed after the polarity of each phase is determined. For example, the polarities of three 1φ transformers or the windings of each of the three legs of a 3φ transformer have been tested and found to be additive. The voltage vectors H1–H2, H2–H3, and H3–H1, representing the induced voltages of transformers A, B, and C are drawn 120° out-of-phase with each other to form a closed delta connection. See Figure 7-38.

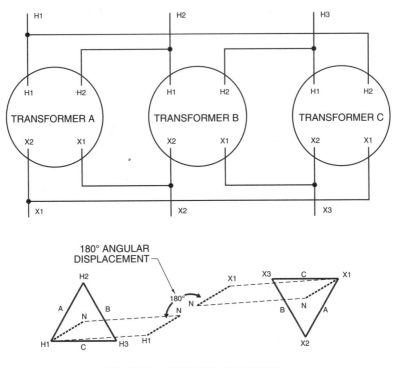

VOLTAGE-VECTOR DIAGRAM

Figure 7-38. Three 1φ transformers of additive polarity may be connected in a delta-to-delta connection with 180° angular displacement.

The low-voltage vector of phase A (full-phase voltage vector X1–X2) is opposite in direction to vector H1–H2 because the polarity is additive. Likewise, the full-phase voltage vector X2–X3, which is the low-voltage vector of phase B, is opposite to that of vector H2–H3. The voltage vector X3–X1, which is the low-voltage vector of phase C, is opposite to vector H3–H1. Therefore, the low-voltage vector diagram is constructed by drawing vector X1–X2 opposite in direction to vectors H1–H2, X2–X3, which is opposite in direction to vector H2–H3, followed by X3–X1 which is opposite in direction to vector H3–H1. This forms a closed delta and is therefore drawn correctly.

If the delta connections forming the 3φ connection on the low-voltage side were improperly made, the voltage vector diagram would not close. For example, if the X1–X2 leads of phase B were interchanged in forming the delta, the voltage-vector diagram would not close. In this case, an unbalanced voltage equal to twice the voltage of one phase would exist in the delta formed and an enormous current would flow through this short-circuited connection.

Voltage-Vector Diagram Test Procedure

An ANSI committee has prepared a test code for transformers. The polarity of each phase may be determined in the same manner as for 1ϕ transformers. The voltage-vector diagrams can be constructed and checked with the ANSI tables for 3ϕ and 6ϕ operation. See Figure 7-39. For example, assume that the 3ϕ connections are delta-to-delta with 180° angular-phase displacement as per Group 2 of 3ϕ transformers without taps. A 3ϕ voltage of low value is applied to leads H1, H2, and H3. With leads H1 and X1 connected, voltage readings are taken between H2 and X2, H3 and X3, H1 and H2, and H2 and X3. For the proper connections and lead markings, the voltage between H2–X2 and H3–X3 should be equal. The voltage for leads H2–X2 should be greater than H1–H2, and the voltage leads for H2–X3 should be less than H2–X2.

The voltage-vector diagram constructed to scale from the voltage measurements should coincide with the diagram for the delta-to-delta-connected transformer. This test is a check for both angular-phase displacement and phase sequence. This method of measurement becomes quite difficult when the high voltage is very large compared to the low voltage. The 3ϕ test, like the 1ϕ test, is limited to transformers having a ratio of transformation of 30:1 or less.

If the internal connections of a 3ϕ transformer are unknown and inaccessible, the polarity test may be omitted. The angular-phase displacement and phase sequence can be determined from test, although it may be necessary to take a greater number of readings. The procedure is the same as described previously. Voltage measurements are made as indicated for each of the angular-phase displacement groups, and the voltage-vector diagram is constructed for check measurement. If none of the diagrams are consistent with the ANSI Test Code for Transformers, the lead markings on the low-voltage side are incorrect. The lettering of two of the low-voltage leads should be interchanged and the voltage measurements repeated. A new voltage-vector diagram is constructed for check measurement. This procedure must be continued until a measurement is found which agrees with the diagram for check measurement.

3φ TRANSFORMER CONNECTION LEAD MARKINGS

GROUP	ANGULAR DISPLACEMENT	DIAGRAM FOR CHECK MEASUREMENT	CHECK MEASUREMENTS
1 ANGULAR DISPLACEMENT 0°	DELTA-DELTA CONN Y-Y CONN	 	CONNECT H1 TO X1 MEASURE H2X2, H3X3 H1H2, H2X3 VOLTAGE RELATIONS (1) H2X2 = H3X3 (2) H2X2 < H1H2 (3) H2X2 < H3X3
2 ANGULAR DISPLACEMENT 180°	DELT-DELTA CONN Y-Y CONN	 	CONNECT H1 TO X1 MEASURE H2X2, H3X3 H1H2, H2X3 VOLTAGE RELATIONS (1) H2X2 = H3X3 (2) H2X2 > H1H2 (3) H2X3 < H2X2
3 ANGULAR DISPLACEMENT 30°	DELTA-Y CONN Y-DELTA CONN	 	CONNECT H1 TO X1 MEASURE H3X2, H3X3 H1H3, H2X2, H2X3 VOLTAGE RELATIONS (1) H3X2 = H3X3 (2) H3X2 < H1H3 (3) H2X2 < H2X3

3φ TRANSFORMERS WITH TAPS

3 ANGULAR DISPLACEMENT 30°	

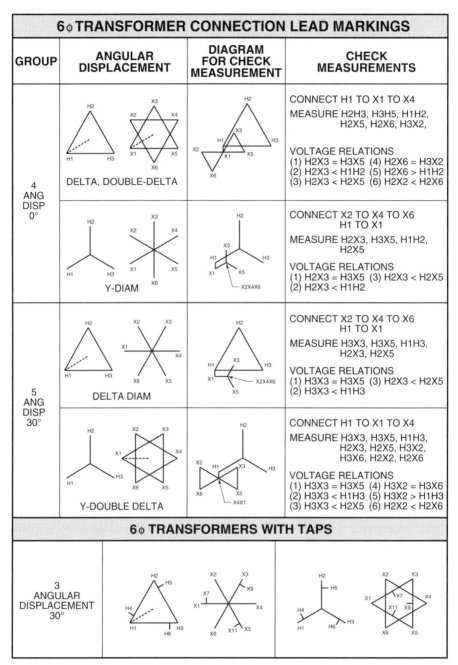

Figure 7-39. ANSI has developed test codes for transformers to help construct and check voltage-vector diagrams for 3φ and 6φ operation.

After a few voltage measurements are taken and voltage-vector diagrams are constructed to scale from the measurements, the correct lead lettering to obtain the proper diagram is apparent and the number of voltage measurements is greatly reduced. These tests do not differentiate between the different connections of any one group of the 3φ connections. For example, in Groups 1 and 2, the voltage relationships for wye-to-wye-connected transformers are the same as for delta-to-delta-connected transformers of the same group. In Group 3, the voltage relationships for delta-to-wye-connected transformers are the same as for wye-to-delta-connected transformers. Therefore, a voltage-vector diagram may be drawn correctly for a transformer which is connected in a wye-to-wye connection, although the sketch seems to indicate that the transformer is connected in a delta-to-delta connection. This differentiation is not necessary as far as parallel operation is concerned. The test does indicate the angular-phase displacement and phase sequence because it indicates the proper grouping of the transformer.

Transformers of the same group may be operated in parallel regardless of their internal connections, provided similarly marked leads are connected and their ratios, voltages, resistances, and reactances permit parallel operation. For example, if all these factors are the same, a delta-to-delta-connected transformer of Group 1 may be operated in parallel with a wye-to-wye-connected transformer of the same group. The same applies to transformers of the other groups.

Transformers of different groups cannot be connected in parallel because of the difference in angular displacement. For example, a delta-to-delta-connected transformer of Group 1 cannot be connected in parallel with a delta-to-delta-connected transformer of Group 2 because there is a difference of 180° in their angular displacement. No interchanging of the external leads changes this displacement, although the order of lettering of the leads may change. To make two such units satisfactory for parallel operation, it is necessary to change the internal delta connections.

When voltage-vector diagrams of transformers which are to operate in parallel are available, the diagrams must coincide and the corresponding terminals must be connected. When the voltage-vector diagrams coincide, leads which are to be connected have the same potential. This is the basic requirement for parallel connections.

Transformer Connections

Tables have been developed to show the proper connections and lead lettering of 1φ transformers of the same or different polarity for each of the five groups of transformers without taps. See Appendix.

Any one connection of any one figure may be connected in parallel with any one connection of another figure by connecting similarly marked leads together, provided the two figures are of the same group and that their ratios, voltages, resistances, and reactances permit parallel operation.For example, any delta-to-delta bank connection may be connected in parallel with any wye-to-wye bank connection by connecting similarly marked leads together, provided the other conditions are the same.

Parallel Operation

Parallel operation of transformers is the connection of two or more transformers which carry a common load. When a transformer has insufficient capacity to deliver a particular load, it may be taken out of the circuit and replaced with a larger unit, or an additional unit may be added to the circuit. An additional unit is added by connecting its primary side to the same source of supply of the original transformer and its secondary side to the same load circuit. The second unit operates in parallel with the first unit. When making this connection, similarly marked leads of both transformers must be connected to the same sides of each circuit. Normally, the parallel connections consist of joining common primary and secondary leads and then connecting the bank to the primary and secondary circuits.

The resistance and reactance of the leads from the transformer to the common points of connection must be included in the resistance and reactance of the transformer. The division of the load between the two parallel-connected transformers may be quite different.

In a satisfactory parallel operation, the transformers connected in parallel share the common load approximately in proportion to their ratings. The most satisfactory condition is obtained when the units divide the common load exactly in proportion to their ratings. For example, if two 50 kVA transformers are connected in parallel with a third 100 kVA transformer, and the common load supplied by the parallel combination is 150 kVA, the two 50 kVA units should each supply $37\frac{1}{2}$ kVA, and the

100 kVA unit should supply 75 kVA. If the common load is 200 kVA, which is the maximum bank capacity of the three transformers obtainable without exceeding the sum of the individual ratings, the two transformers rated at 50 kVA should each supply 50 kVA, and the 100 kVA unit should supply 100 kVA to the common load. In both cases, the division of load between the three transformers is proportional to their rating.

Satisfactory Parallel Operation Requirements

The requirements for satisfactory operation of two or more transformers connected in parallel include the same polarity or angular-phase displacement and phase sequence, voltage ratio, and percent impedance. The ratio of the resistance component to the reactance component of each impedance should also be the same.

Displacement. In a 1ϕ circuit, the secondary windings of two transformers connected in parallel are either in the same time-phase relationship (the two voltages have 0° displacement between each other), or are in an opposite time-phase relationship (the two voltages have 180° displacement between each other). The parallel connection of the two windings constitutes a closed series circuit. If the two voltages are of the same time-phase relationship, the induced voltages from either common connection to the other are in the same direction (opposing each other) and no current flows in the series circuit. With no load connected across the terminals of the parallel connection, satisfactory paralleling conditions are established if no current flows in the series circuit. The two windings are of the same polarity.

If the two voltages are in time-phase opposition, the induced voltages from either common connection to the other are in opposite directions. The voltages add to one another in the series circuit and a large current flows which is equal to a short circuit on both transformers. The two windings are of opposite polarity and the parallel connection formed is unsatisfactory.

The polarity designation must not be confused with the terms additive and subtractive polarity as applied to the individual units. Additive and subtractive polarity indicate the relative directions of the induced voltages and are given to assist in making the proper parallel connections. If the leads are properly marked with respect to the definitions for additive and subtractive polarity, similarly marked leads are connected to obtain the proper parallel connection.

For example, three separate banks, each consisting of two 1φ transformers, are connected in parallel. See Figure 7-40. The connections contain two transformers of additive polarity connected in parallel, two transformers of subtractive polarity connected in parallel, and a transformer of additive polarity connected in parallel with a transformer of subtractive polarity. In each of the connections, the X1 leads of each transformer are connected and the two X2 leads are connected. The induced voltages in each connection from X1 to X2 are the same through each of the transformers in that the lead marking is determined by the direction of the induced voltages. In each parallel connection, the two transformers are of the same polarity.

If lead markings are unknown when making a parallel connection, the two leads located in similar positions with respect to the bank are first connected together, and a low-current rated fuse is inserted between the remaining two leads. The voltage across the fuse is zero and no current flows if the windings are of the same polarity. If the windings are of opposite polarity, the voltage across the fuse is equal to twice the voltage of either winding and an equivalent short-circuit current

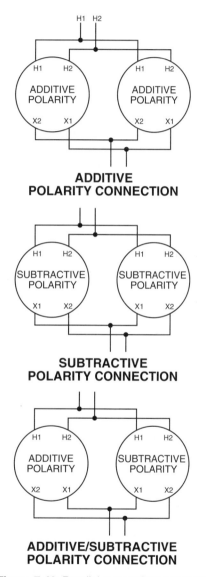

Figure 7-40. Parallel connections normally consist of joining common primary and secondary leads and then connecting the bank to the primary and secondary circuits.

flows and blows the fuse. To obtain satisfactory operation in this case, it is necessary to interchange the connections of one of the units. This is equivalent to reversing the direction of the induced voltage of one unit, making both the same polarity.

Polarity may be neglected entirely when connecting 3ϕ transformers of the same voltage ratio in parallel. For satisfactory parallel connections, the angular-phase displacement and phase rotation between the two units to be connected in parallel must be the same. For example, a wye-to-wye-connected transformer having an angular-phase displacement of 0° cannot be connected in parallel with a wye-to-delta-connected transformer having an angular-phase displacement of 30°. The secondary voltages of the two connections would be 30° out-of-phase with each other. Because the voltages are out-of-phase with each other, if connected in parallel, a voltage exists within the parallel connection, sending current through each of the two transformers. This results in unsatisfactory parallel operation. Similarly, if the phase rotation of the secondaries of the two 3ϕ transformers are in opposite directions, a voltage exists within the parallel connection. The voltage sends current through the two transformers, resulting in unsatisfactory parallel operation.

To simplify the parallel connection of 3ϕ transformers without having to test for polarity, angular-phase displacement, and phase rotation, the lead markings have been standardized and the various 3ϕ connections have been placed in three different groups, depending on their angular displacement. All transformers of any one group may be connected in parallel with each other. A transformer of one group cannot be connected in parallel with a transformer of another group because of the difference in angular displacement. Transformers from different groups cannot be connected in parallel by interchanging the external leads. To change the angular-phase displacement from 0° to 180° or vice versa, the internal connections of the coils forming the wye or delta connection must be changed.

Voltage Ratio. If the voltage ratio of two transformers connected in parallel is not the same, the difference in voltage between the two windings causes a current to flow within the parallel circuit at all times. Considering the parallel connection comprising two windings connected in series, the voltage difference sends a circulating current through the two windings. The amount of circulating current is limited by the sum of the impedances of the two transformers. If the impedance voltage of the two transformers is each 5%, 5% of normal voltage impressed on one winding circulates full-load current when the other winding is short-circuited. The total impedance of the two windings in series is 10% when the two transformers are connected in parallel.

To limit the circulating current to 10% of normal value, the voltage producing the current flow could not exceed 1% of the voltage of the winding in terms of which it is expressed. For example, if the secondary voltage of one of the transformers is 120 V, the voltage of the other transformer must be within 1.2 V (1% of 120 V) above or below 120 V to limit the circulating current to a value of 10% of the normal full-load rating. This assumes both units have the same capacity rating. The circulating current, which is at a 90° angle to the voltage, flows at no load as well as at full load. Therefore, when an external load is placed on the two units connected in parallel, the circulating current adds vectorially to the load current in one winding and subtracts vectorially from the load current in the other. To calculate the amount of circulating current, apply the formula:

$$\%I_c = \frac{100 \times \% \, e}{\%IZ_1 + \%IZ_2}$$

where

$\%I_c$ = circulating current (in percent of the normal load current of one transformer)

$\%e$ = difference in voltage between the two windings expressed as a percent of normal voltage

$\%IZ_1$ = percent impedance of unit No. 1

$\%IZ_2$ = percent impedance of unit No. 2

This formula assumes that the capacity of both units is the same. If the capacities are different, the percent impedances in the formula should be based on the same kVA rating, and the percent circulating current is then a percent of the normal load current at the same kVA rating.

Example: Calculating Circulating Current

What is the circulating current when a 50 kVA transformer having a secondary voltage of 100 V and an impedance of 4% is placed in parallel with a 100 kVA transformer having a secondary voltage of 102 V, and an impedance of 4%? *Note:* The impedance of the 100 kVA transformer at 50 kVA is 2% (e = 2 V).

$$\%I_c = \frac{100 \times \% e}{\%IZ_1 + \%IZ_2}$$

$$\%I_c = \frac{100 \times 2}{4 + 2}$$

$$\%I_c = \frac{200}{6}$$

$$\%I_c = \mathbf{33.33\%}$$

The circulating current, which is expressed as a percent of the normal load current of the 50 kVA, 100 V transformer is 33.33%. The normal load current is 500 A (50,000 ÷ 100 = 500 A). Therefore, the circulating current is 166.7 ($33\frac{1}{3}\%$ of 500 = 166.7 A).

A small difference between the ratios of transformers connected in parallel produces a relatively large circulating current. For satisfactory parallel operation, the circulating current should not exceed 10% of the normal load current.

Two transformers of different voltage ratios can be connected in parallel by the use of a balance coil. See Figure 7-41. A *balance coil* is a single-winding transformer having a tap located at some point between its two ends.

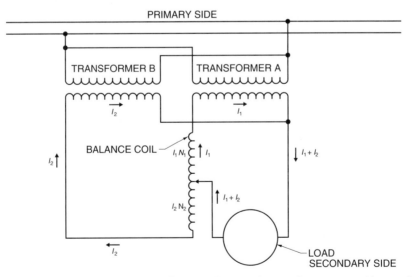

Figure 7-41. Two transformers of different voltage ratios can be connected in parallel by the use of a balance coil.

The position of the tap is determined by the desired division of load on the two transformers connected in parallel. The balance coil receives its excitation from the voltage difference of the two windings connected in parallel. For this reason, the current which normally flows in the two windings, due to the voltage difference, is reduced to an amount equal to the magnetizing current of the balance coil.

The tap is located so that when the desired division of current through each of the windings is obtained, the ampere-turns on one side of the tap are equal to the ampere-turns on the other side. Because the ampere-turns on the two sides are in opposition (as in an autotransformer), they neutralize each other with respect to the magnetizing effect of the magnetic circuit. A 3ϕ balance coil may similarly be used to connect 3ϕ transformers in parallel.

Percent Impedance. The division of the external load between transformers connected in parallel is inversely proportional to their ohmic impedances. The load divides among the several transformer units until the terminal voltages of all of them have the same value.

The unit, which maintains a higher voltage (has a lower impedance), takes more current until its terminal voltage falls to a value equal to the terminal voltage of the other unit, or the currents through the several units in parallel are such that the impedance drops through the several units are the same. Therefore, the loads on the transformers connected in parallel vary inversely as the impedances of the transformers (expressed in ohms). If the ratings of the two transformers are the same, the loads divide inversely as the impedances of the transformers expressed in percent of normal voltage. To calculate the division of load between any number of 1ϕ transformers operating in parallel, apply the formula:

$$kVA_1 = \frac{\left(\dfrac{kVA}{\%IZ}\right)_1}{\left(\dfrac{kVA}{\%IZ}\right)_1 + \left(\dfrac{kVA}{\%IZ}\right)_2 + \ldots} \times kVA_L$$

where

kVA_1 = kVA load supplied by Transformer 1

$\left(\dfrac{kVA}{\%IZ}\right)_1$ = kVA rating of Transformer 1 divided by its percent impedance

$\left(\dfrac{kVA}{\%IZ}\right)_2$ = kVA rating of transformer 2 divided by its percent impedance

kVA_L = total kVA of the connected load

Example: Calculating Transformer Load

What is the portion of the total load supplied by each transformer if a 50 kVA, 2400:240 V transformer having an impedance of 4% is connected in parallel with a 75 kVA, 2400:240 V transformer having an impedance of 6%. *Note:* The transformers supply a load of 125 kVA.

To calculate the portion of the total load supplied by the 50 kVA transformer, apply the formula:

$$kVA_1 = \dfrac{\left(\dfrac{kVA}{\%IZ}\right)_1}{\left(\dfrac{kVA}{\%IZ}\right)_1 + \left(\dfrac{kVA}{\%IZ}\right)_2} \times kVA_L$$

$$kVA_1 = \dfrac{\dfrac{50}{4}}{\left(\dfrac{50}{4}\right) + \left(\dfrac{75}{6}\right)} \times 125$$

$$kVA_1 = \left(\dfrac{12.5}{12.5 + 12.5}\right) \times 125$$

$$kVA_1 = \left(\dfrac{12.5}{25}\right) \times 125$$

$$kVA_1 = .5 \times 125$$

$$kVA_1 = \textbf{62.5 kVA}$$

To calculate the portion of the total load supplied by the 75 kVA transformer, apply the formula:

$$kVA_2 = \frac{\left(\dfrac{kVA}{\%IZ}\right)_2}{\left(\dfrac{kVA}{\%IZ}\right)_1 + \left(\dfrac{kVA}{\%IZ}\right)_2} \times kVA_L$$

$$kVA_2 = \frac{\dfrac{75}{6}}{\left(\dfrac{50}{4}\right) + \left(\dfrac{75}{6}\right)} \times 125$$

$$kVA_2 = \left(\frac{12.5}{12.5 + 12.5}\right) \times 125$$

$$kVA_2 = \left(\frac{12.5}{25}\right) \times 125$$

$$kVA_2 = .5 \times 125$$

$$kVA_2 = \textbf{62.5 kVA}$$

The portion of the total load supplied by the 75 kVA transformer is 62.5 kVA.

The 50 kVA transformer is overloaded and the 75 kVA unit is operating below its capacity. If the load on the 50 kVA unit is to be limited to 50 kVA, the external load of the bank must be reduced in the ratio of 62.5 to 50 or from 125 kVA to 100 kVA. The load on the 75 kVA unit must be reduced by the same ratio and would be 50 kVA also.

The formulas are correct only when the ratio of resistance to reactance of each of the transformers is the same. In this case, the current in each of the transformers is in phase. If the ratios are not the same, the current in the transformers are not in phase and the sum of the winding current is greater than the line current. The phase angle of the current in the transformer is equal to the angle in which the cosine is the percent reactance divided by the percent impedance. If the difference in phase angle between two transformers connected in parallel is less than 15°, the line current may be assumed to be the arithmetical sum of the winding currents, as found from the formulas without appreciable error.

The percent impedances of all transformers should be the same for satisfactory parallel operation. If this is the case, neglecting the error introduced if their resistances and reactances are not of the same ratio, the load divides in the individual units in the ratio of their capacities, and the total permissible load equals the sum of the capacities of the individual units connected in parallel.

It is possible, but not practical or economical to obtain such a division of load with units of unlike impedance. This requires installing an impedance coil in the form of a reactor in series with the transformer having the lower impedance. Although seldom used, the reactance of this reactor should be such that when placed in series with the low-impedance unit, the percent impedance is raised to a value equal to that of the other transformer in the parallel circuit.

A dry-type reactor may be connected in series with the low-voltage secondary side of a distribution transformer to raise the percent impedance of one transformer to that of another. See Figure 7-42. This ensures that the total load is divided between the two transformers in proportion to their kVA ratings. The increased impedance required in the circuit is obtained by connecting a reactor in series with the low-voltage leads. A reactor of this type or of similar construction is suitable for voltages up to 600 V. For voltages above 34,500 V, it becomes more economical to use an oil-insulated reactor.

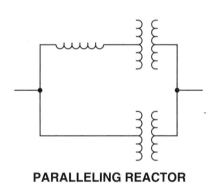

PARALLELING REACTOR

Figure 7-42. A dry-type reactor may be connected in series with the low-voltage secondary side of a distribution transformer to raise the percent impedance of one transformer to that of another.

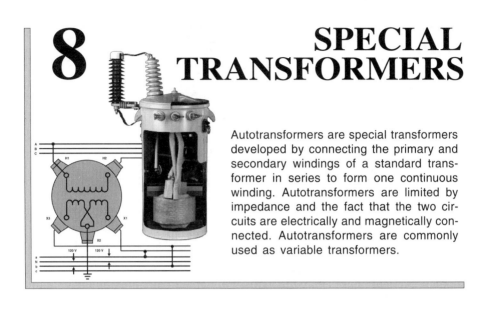

8 SPECIAL TRANSFORMERS

Autotransformers are special transformers developed by connecting the primary and secondary windings of a standard transformer in series to form one continuous winding. Autotransformers are limited by impedance and the fact that the two circuits are electrically and magnetically connected. Autotransformers are commonly used as variable transformers.

AUTOTRANSFORMERS

Transformers normally contain two independent circuits. The cicuits are the primary circuit, which receives the energy, and the secondary circuit, which delivers the energy to the load circuit. See Figure 8-1. In this transformer, a secondary load current of I_2 amps at a voltage of E_2 volts is supplied by the primary winding having a current of I_1 amps when connected across a voltage source of E_1 volts. Neglecting the losses in the transformer, the power transformed is equal to $I_2 \times E_2$ volt-amperes which is equal to the input power $I_1 \times E_1$ volt-amperes.

An autotransformer is developed by connecting the two windings of a transformer in series to form one continuous winding. See Figure 8-2. An *autotransformer* is a transformer in which at least two windings have a common section. An autotransformer is a self-induced transformer. If a voltage is impressed across the connected windings, current may be taken from the transformer between the point where the two windings are connected and either of the two ends of the winding. An autotransformer is any transformer in which the primary and secondary circuits have a portion of the winding in common.

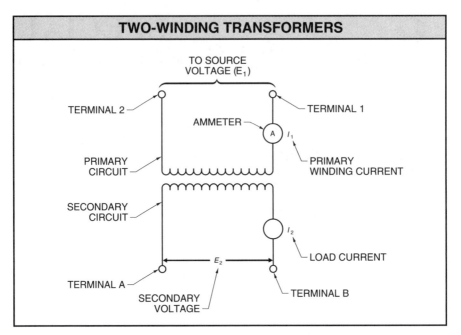

Figure 8-1. Transformers normally contain independent primary and secondary circuits.

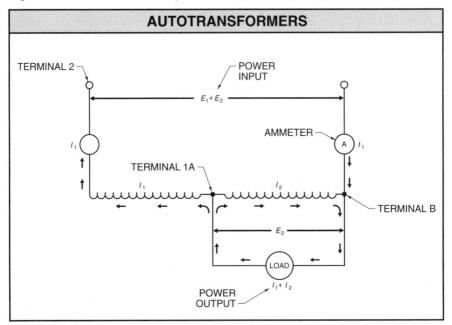

Figure 8-2. An autotransformer is developed by connecting the two windings of a transformer in series to form one continuous winding.

The electrical characteristics of a two-winding transformer change when connected as an autotransformer. For example, the voltage rating of the primary winding of an autotransformer is the sum of the primary and secondary voltages of a two-winding transformer with identical characteristics. The external load current is the sum of the primary and secondary currents of a two-winding transformer with identical characteristics. The power input of an autotransformer is $I_1 \times (E_1 + E_2)$ volt-amperes, which is equal to the power output of $(I_1 + I_2) \times E_2$ volt-amperes. The output is increased by $I_1 \times E_2$ volt-amperes because of the connected windings.

In this autotransformer, a voltage impressed on terminal 2 and terminal B induces a counter EMF equal to the impressed voltage in the winding. When a load is connected to terminal 1A and terminal B, the counter electromotive force between these terminals sends current through the secondary. The current in the two windings are the same as when the windings are connected as a two-winding transformer. Therefore, the power transformed by the unit in either case is equal to $I_2 \times E_2$ volt-amperes. The remainder of the load, equal to $I_1 \times E_2$ volt-amperes, is not transformed but is conducted directly from the primary winding to the load.

The portion of the winding from terminal 2 to terminal 1A may be considered the primary winding, and the portion of the winding from terminal 1A to terminal B may be considered the secondary winding. The volt-amperes of the primary winding is equal to the volt-amperes of the secondary winding, just as in the case of a two-winding transformer. Because part of the secondary load in the autotransformer is transferred rather than transformed, the output when connected as an autotransformer is greater than when connected as a two-winding transformer.

Output Rating

Less material is required in an autotransformer than in a two-winding transformer for a given output because a single winding in an autotransformer serves as both the primary and secondary circuits. The output of an autotransformer is greater with the same amount of material. For example, a two-winding transformer is rated at 10 kVA (10,000 VA) with a primary voltage (E_1) of 1000 V and a secondary voltage (E_2) of 100 V. The primary current (I_1) is 10 A and the secondary current (I_2) is 100 A.

When connected as an autotransformer, the primary voltage ($E_1 + E_2$) is 1100 V and the secondary voltage (E_2) is 100 V. Assuming the same currents in the winding (I_1 = 10 A and I_2 = 100 A), the secondary load current is 110 A ($I_1 + I_2$ = 110 A) and the total output is 11,000 VA (110 × 100 = 11,000 VA or 11 kVA).

The ratio of transformation as a two-winding transformer is 10:1 ($E_1 \div E_2$ = 1000 ÷ 100 = 10:1). The ratio of transformation as an autotransformer is 11:1 [($E_1 + E_2$) ÷ E_2 = 1100 ÷ 100 = 11:1]. The ratio of transformation increased 10%. Likewise, the output is increased from 10 kVA to 11 kVA or 10%. The rating of this transformer when operating as an autotransformer is 11 kVA.

The output rating of an autotransformer may be expressed in terms of the ratio of its voltage transformation and the rating of the same winding when operated as a two-winding transformer. To calculate the output rating of an autotransformer, apply the formula:

$$A_r = Tr \times \frac{1}{1 - \frac{1}{R}}$$

where

A_r = autotransformer rating (in VA)

T_r = two-winding transformer rating (in VA)

R = autotransformer voltage ratio

Example: Calculating Autotransformer Output Rating

What is the output rating of an autotransformer connected from a two-winding transformer with a rating of 10,000 VA and a high-voltage to low-voltage ratio of 11:1?

$$A_r = Tr \times \frac{1}{1 - \frac{1}{R}}$$

$$A_r = 10,000 \times \frac{1}{1 - \frac{1}{11}}$$

$$A_r = 10,000 \times \frac{11}{10}$$

$$A_r = \textbf{11,000 VA (11 kVA)}$$

This expression may be rearranged to give the kilovolt-amperage of transformer parts required to make an autotransformer of a specified output rating and voltage ratio. The rearranged formula used to calculate the kilovolt-amperage of transformer parts required to make an autotransformer of a specified output and voltage ratio is $T_r = A_r \times 1 - (1 \div R)$.

For example, an autotransformer is required with a rating of 11 kVA and a ratio of transformation of 11:1. The kVA of transformer parts equals 10 kVA [$11 \times 1 - (1 \div 11) = 11 \times {}^{10}\!/_{11} = 10$ kVA].

By using an autotransformer, an output of 11 kVA can be obtained from the same parts that would be required to give an output of only 10 kVA when connected as a two-winding transformer. As the ratio of high voltage to low voltage of the autotransformer decreases, the parts necessary to produce the autotransformer capacity becomes smaller.

The quantity $1 - (1 \div R)$ approaches 1 when the voltage ratio becomes very large. In this case, the kilovolt-amperage of transformer parts is nearly equal to the rating of the transformer. There is no advantage in connecting the unit as an autotransformer.

As the voltage ratio becomes smaller and approaches 1, the quantity $1 - (1 \div R)$ approaches 0. This means that as the secondary voltage of the autotransformer approaches the primary voltage value, the parts required to build such a unit of a given kilovolt-ampere rating becomes smaller and smaller. The reduction in size as the ratio of transformation decreases is one of the main reasons for the use of an autotransformer in place of a two-winding transformer.

For example, a 150 kVA load must be stepped up from 220 V to 250 V by an autotransformer. The two-winding transformer must have a rating of 150 kVA if the change in voltage is obtained with a two-winding transformer. The ratio of transformation of the high voltage to low voltage is $250 \div 220$. Then $1 \div R = 250 \div 220 = .88$ and $1 - (1 \div R) = 1 - .88 = .12$.

The kilovolt-amperes of the transformer parts required to transform 150 kVA from 220 V to 250 V equals 18 kVA ($150 \times .12 = 18$ kVA). This means that an autotransformer required to transform 150 kVA from 220 V to 250 V is no larger than an 18 kVA two-winding transformer having primary and secondary windings of 220 V and 30 V respectively.

As the voltage ratio increases, the economic advantage of an autotransformer over a two-winding transformer decreases. For this reason, autotransformers are rarely made with a ratio greater than 4:1 or 5:1. This is because the material saved in a high-ratio autotransformer is relatively small. Also, the fact that the high- and low-voltage windings are electrically connected requires, if the circuit is ungrounded, that both have the same insulation to ground. This adds considerable cost to a high-ratio autotransformer. A two-winding transformer is better for such applications.

Autotransformer Limitations

Autotransformers are limited by impedance and the fact that the two circuits are electrically and magnetically connected. The impedance of an autotransformer is much less than the impedance of a two-winding transformer. For example, if a two-winding transformer has an impedance of about 3% of rated primary voltage as an autotransformer, its impedance is equal to 3% times the difference between the primary and secondary voltages over the primary voltage. This occurs because the characteristics are based on the size of transformer parts.

By reducing the impedance, a much higher current flows when a short-circuit fault occurs. A means is required to make the transformer self-protecting should such a fault occur, because transformers are not built to withstand the high short-circuit fault current. Self-protection is accomplished by providing an external impedance. External impedance is provided by other transformers in the circuit, reactors, other devices, and in some cases, within the actual power lines.

The other limitation of an autotransformer is that the two circuits are electrically and magnetically connected. It is possible that the section of the winding, common to the primary and secondary, may become open. In such a case, anyone coming in contact with the secondary is subject to the primary voltage, or the primary may become grounded and establish a high voltage between one of the low-voltage conductors and ground.

Autotransformer Applications

Autotransformers are used in starting rotating apparatus such as synchronous and induction motors. A synchronous motor is started at about one-third of normal voltage. An induction motor is started at about two-thirds of running voltage. The fact that the primary and secondary voltages of the autotransformer are not insulated from each other is not objectionable. This is because the voltages of such machines are relatively low and because a starting autotransformer is connected to the line only during the starting cycle. The size of the autotransformer is further decreased because the starting cycle is short (usually only 1 minute). The materials can be worked at much higher densities without exceeding a safe temperature rise.

Starting autotransformers are usually connected in an open-delta or wye configuration. An open-delta configuration is used for small motors and the wye configuration is used for large machines. The wye configuration gives a better voltage balance than the delta configuration. The open-delta configuration provides a simple means of voltage control, especially when more than one starting voltage is required.

Transformers connected in an open-delta configuration require the changing of taps in two coils only. Transformers connected in a wye configuration require the changing of taps in three coils. Autotransformers for starting duty may be 1ϕ units connected in open-delta or wye configurations, or they may be 3ϕ units. Three-phase units consist of two windings on a 3ϕ magnetic circuit. Three-phase units are preferred because they require less space than three 1ϕ units of the same capacity.

Autotransformers for power purposes are used when the primary and secondary voltages do not differ greatly, such as 115 kV and 69 kV and where greatest economy results. Such autotransformers are usually connected in a wye configuration with the neutral grounded. See Figure 8-3. Such transformers are usually provided with a delta-connected tertiary winding to provide a path for the flow of the third harmonic of exciting current. This current is usually very small and the delta winding added has a small percentage of the capacity of the main autotransformer winding. The delta winding is sometimes increased in size and used for supplying power to a synchronous condenser for power factor correction.

While a 3ϕ transformation may be obtained with three 1ϕ autotransformers or one 3ϕ autotransformer connected in a delta configuration, they are seldom used because they are very inefficient. Furthermore, a voltage transformation greater than 2:1 cannot be obtained. Delta-connected autotransformers also cause a phase shift of the secondary voltage as compared with wye-connected autotransformers. The phase displacement also varies for each different ratio of transformation. Therefore, it is impossible to parallel wye-connected and delta-connected autotransformers. See Figure 8-4.

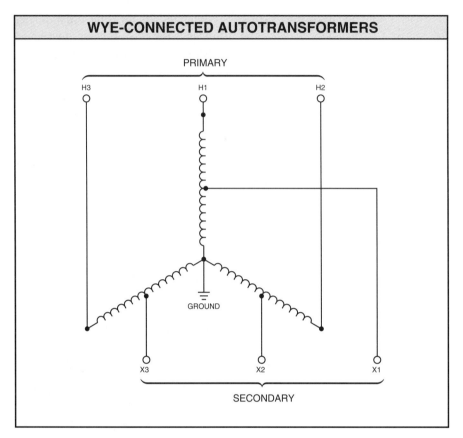

WYE-CONNECTED AUTOTRANSFORMERS

PRIMARY

H3   H1   H2

GROUND

X3   X2   X1

SECONDARY

Figure 8-3. Autotransformers for power purposes are usually connected in a wye configuration with the neutral grounded.

Autotransformers For Miscellaneous Uses

Autotransformers may be used as a preventive autotransformer in bridging across successive taps in a power transformer in which the voltage ratio is changed while the transformer is carrying a load. The National Electrical Code® (NEC®) permits the use of autotransformers (or equipment, such as buck and boost transformers connected as autotransformers) on existing installations without requiring a common primary and secondary grounded conductor if the transformation is 208 V to 240 V or 240 V to 208 V. See Figure 8-5. Thus, an autotransformer can be used for voltage regulation.

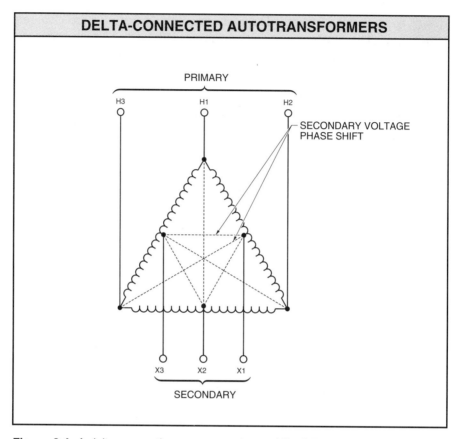

Figure 8-4. A delta connection causes a phase shift of the secondary voltage.

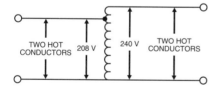

Figure 8-5. The NEC® permits the use of autotransformers on existing installations without requiring a common primary and secondary grounded conductor if the transformation is 208 V to 240 V or 240 V to 208 V.

Autotransformers are commonly used as variable transformers. See Figure 8-6. A *variable transformer* is a continuously adjustable autotransformer consisting of a single layer of wire wound on a toroidal core and a carbon brush, which traverses this winding. A *toroidal core* is a donut-shaped core. The brush track is made by removing a portion of the insulation from each turn of the winding, forming a series of commutator elements. The basic principle is that of a tap changing transformer. The brush is always in contact with one or more wires and continuously taps off any desired fraction of the winding voltage. It is possible to remove the contact under load without interrupting the circuit.

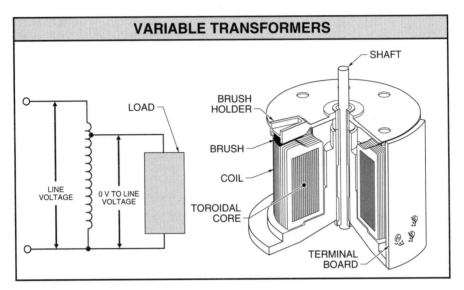

Figure 8-6. Autotransformers are commonly used as variable transformers.

9

REACTORS

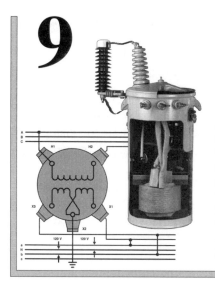

A reactor introduces reactance into a circuit. A current-limiting reactor is a reactor which limits the current that can flow in a circuit under short-circuit conditions. A saturable-core reactor is a magnetic-core reactor in which the reactance is controlled by changing the saturation of the core through variation of a superimposed flux. A shunt reactor is a reactor that is connected in shunt to an electrical system.

CURRENT-LIMITING REACTORS

A *reactor* is a device that introduces reactance into a circuit. A *current-limiting reactor* is a reactor which limits the current that can flow in a circuit under short-circuit conditions or under other operating conditions, such as starting, synchronizing, etc.

Inductance

Inductance is the property of an electrical circuit by which a varying current induces an electromotive force in the circuit or in a neighboring circuit. A magnetic field is established around and at right angles to a conductor whenever current flows in the conductor. The strength of the magnetic field is directly proportional to the quantity of electricity flowing. The magnetic field increases as the current increases. The magnetic field decreases as the current decreases.

The magnetic field (flux lines) originates from the center of the current-carrying conductor and expands or contracts in increasing or decreasing circles as the current increases or decreases. A counter electromotive force is induced in the conductor as the flux lines cut the conductor. The circuit has an inductance of 1 H (henry) when 1 V is induced in the circuit by a uniform change of the current at a rate of 1 A per second.

Reactance

Reactance (*X*) is the measure of the induced voltage due to the inductance of a circuit. Reactance is directly proportional to the rate of change of current (frequency) of the circuit. To calculate reactance, apply the formula:

$X = 2\pi fL$

where

X = reactance (in Ω)

2 = constant

π = 3.1416

f = frequency of circuit (in cycles per sec)

L = inductance of circuit (in H)

Example: Calculating Reactance

What is the reactance in a circuit that has a frequency of 60 Hz and an inductance of .002 H?

$X = 2\pi fL$

$X = 2 \times 3.1416 \times 60 \times .002$

$X = .754\ \Omega$

The reactive voltage may be expressed as $2\pi fLI$ or IX where I is the rms (root-mean-square) current in amps. The reactive voltage is normally kept to a low value because of its adverse effect on voltage regulation and the power factor of the system. For example, if a transformer has little or no leakage reactance, the reactive voltage vectors $I'X_p$ and I_sX_s are very small (equal to zero). See Figure 9-1.

Assuming the same induced voltages E_p and E_s on the primary and secondary sides respectively, the impressed voltage E' required to give an induced voltage of E_p volts is greatly reduced. Also, the secondary terminal voltage E'', at a load current of I_s amperes is increased with a resulting improvement in voltage regulation. Likewise, if the load circuit has little or no reactance, the load current I_s is nearly in-phase with the terminal voltage E'' with a resulting improvement in the power factor of the circuit.

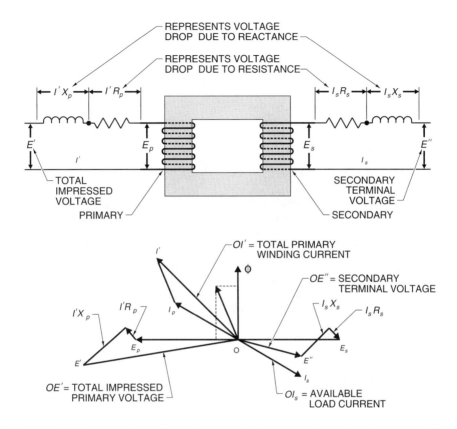

Figure 9-1. The reactive voltage is normally kept to a low value because of its adverse effect on voltage regulation and the power factor of the system.

However, if an AC circuit could be made without reactance, the short-circuit current would be limited only by the resistance of the circuit. This results in a large current flow during a short circuit, especially if there is a large amount of power available for feeding current into the short circuit. The conductor in the circuit must be braced to withstand the stresses set up by the magnetic forces of the large currents, and circuit breakers must be capable of interrupting these currents. For this reason, reactance is commonly added to a circuit so that the short-circuit current is limited to a predetermined value that is safely and economically handled. This additional reactance is provided in current-limiting reactors. See Figure 9-2.

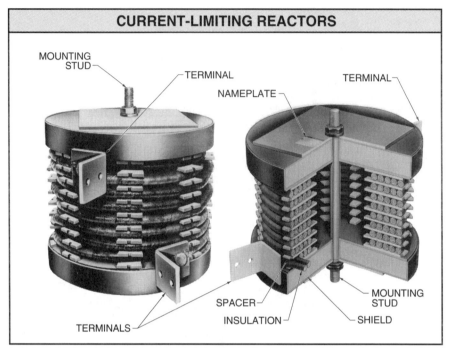

CURRENT-LIMITING REACTORS

MOUNTING STUD

TERMINAL

TERMINAL

NAMEPLATE

MOUNTING STUD

SPACER

INSULATION

SHIELD

TERMINALS

ABB Power T&D Company Inc.

Figure 9-2. A current-limiting reactor limits the current that can flow in a circuit under short-circuit conditions or under other operating conditions, such as starting, synchronizing, etc.

Ratings

A current-limiting reactor may be rated in ohms reactance at a given frequency and current-carrying capacity or by expressing the voltage drop across the coil as a percentage of the normal voltage of the circuit when carrying the given current at a given frequency. The circuit voltage used should be the voltage between the line and neutral when connected in a 3ϕ system. For example, a reactor having an inductance of .002 H has a reactance of .754 Ω at 60 Hz ($2 \times 3.1416 \times 60 \times .002 = .754\ \Omega$). The voltage drop ($E_x$) across the reactor is equal to the ohms reactance multiplied by the current. For example, if 400 A is flowing in the circuit, the voltage drop equals 301.6 V (400 A $\times$ .754 Ω = 301.6 V). If this reactor is used in a 1ϕ circuit of 11,000 V, the percent reactance is 2.742%

($^{301.6 \times 100}/_{11,000}$ = 2.742%). If this reactor is used in an 11,000 V, 3ϕ circuit in which the voltage to neutral is 6350 V (11,000 ÷ $\sqrt{3}$ V = 6350 V), the percent reactance with 400 A flowing in the line is 4.749% ($^{301.6 \times 100}/_{6350}$ = 4.749%).

The short-circuit current through this reactor with full voltage maintained at its terminals can be calculated from the ohms or the percent reactance. The short-circuit current is calculated by dividing the circuit voltage by the ohms reactance or dividing the normal current times 100 by the percent reactance. In the 3ϕ circuit example, the short-circuit current, calculated by dividing the circuit voltage by the ohms reactance, is 8422 A ($^{6350}/_{.754}$ = 8422 A or $^{400 \times 100}/_{4.749}$ = 8422 A). The short-circuit current, calculated by dividing the normal current times 100 by the percent reactance, is 8422 A ($^{400 \times 100}/_{4.749}$ = 8422 A).

Construction

The current range of the reactor in the example is from a normal current of 400 A to a maximum short-circuit current of 8422 A or approximately 21 times its normal full-load rating. The current range of a current-limiting reactor may vary from 4 to 33⅓ times its normal full-load current rating. The ohms reactance of the reactor must be constant throughout the entire range of current variation or the short-circuit current varies.

The magnetic induction in the core increases as the voltage builds up across the reactor if an iron core is used for the magnetic circuit to obtain the necessary inductance at normal current rating. As the current and voltage increase, the saturation point in the core is reached and each succeeding increase in current must be larger than the previous to produce the same increase in voltage. The current increases more rapidly than the voltage, which is equivalent to a corresponding decrease in reactance. To maintain a constant value of reactance, the magnetic circuit of the reactor must be worked at so low an induction point that the saturation point is not reached until after the maximum short-circuit current is obtained.

For current-limiting purposes, it is not economical to use a magnetic circuit of iron with air gaps as is done in the case of a paralleling reactor in which the range of current seldom exceeds 2½ times its normal full-load current rating. The magnetic circuit of a current-limiting reactor is therefore an air core, giving a voltage drop across the reactor at all times directly proportional to the current flowing through it.

The current-carrying circuit of a current-limiting reactor is circular in shape. It consists of several turns of copper cable insulated with cotton, asbestos, or spun glass tape. The turns are wound into discoidal layers. The horizontal spacers between layers provide an extra layer of insulation. These spacers have a hole through which vertical tie rods are passed for bolting the entire structure between concrete disks at the top and bottom. These concrete disks have inserts cast in them for attaching the insulating feet and terminal supports. Cable ends are brazed into the terminal supports and all connections not bolted are either welded or brazed. This form of construction gives maximum strength to resist the mechanical stresses set up by the magnetic field during short-circuit conditions.

Types

Current-limiting reactors may be dry-type or oil-immersed. Dry-type reactors are used indoors for all voltages up to approximately 34,500 V and outdoors for voltages to 25,000 V. Oil-immersed reactors have current-carrying coils that are normally wound in cylindrical layers or pancake form for voltages of 34.5 kV and below and in pancake form for voltages above 34.5 kV.

Dry. Dry-type current-limiting reactors depend entirely on the surrounding air for carrying away the heat developed in the winding. For this reason, adequate space must be provided for a free circulation of air. See Figure 9-3. Care must be taken to keep the reactor at some distance from all magnetic materials, such as steel channels, S beams, plates, and other steel structures because the reactor uses the air as a return path for the flux. The minimum distance is normally specified by the manufacturer. If the reactor is to be mounted in a metal cell, care must be taken to break up electrical and magnetic circuits adjacent to the reactor to eliminate any paths in which the flux from the reactor might induce a voltage sufficient to cause excessive eddy-current flow.

Oil-Immersed. Oil-immersed current-limiting reactors may be used for indoor or outdoor service on any voltage for which a reactor may be required. See Figure 9-4. Oil-immersed reactors differ from dry-type reactors in that a magnetic shield is placed around the outside of the coil to confine the flux within the limits of the shield. This prevents flux from flowing into the wall of the tank in which the reactor is placed.

DRY-TYPE CURRENT-LIMITING REACTORS

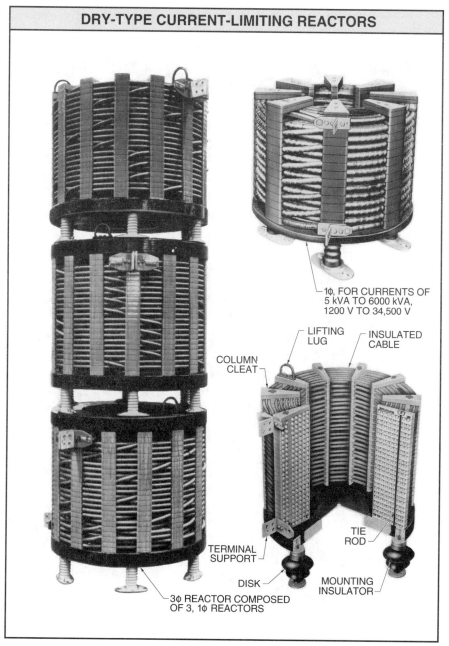

1φ, FOR CURRENTS OF
5 kVA TO 6000 kVA,
1200 V TO 34,500 V

LIFTING LUG

INSULATED CABLE

COLUMN CLEAT

TERMINAL SUPPORT

DISK

TIE ROD

MOUNTING INSULATOR

3φ REACTOR COMPOSED OF 3, 1φ REACTORS

ABB Power T&D Company Inc.

Figure 9-3. Dry-type current-limiting reactors depend entirely on the surrounding air for carrying away the heat developed in the winding.

OIL-IMMERSED CURRENT-LIMITING REACTORS

1φ, 11,136 kVA, 60 Hz, 138 kV BANK OF OIL-IMMERSED CURRENT-LIMITING REACTORS

OIL-IMMERSED REACTOR WITH TANK REMOVED

ABB Power T&D Company Inc.

Figure 9-4. Oil-immersed current-limiting reactors may be used for indoor or outdoor service on any voltage for which a reactor may be required.

The magnetic shield is constructed of laminated high-silicon sheet steel which is welded to the tank wall at each end. The heavy steel angles at each end, as well as the hoops around the body of the iron shield, have insulated joints to avoid the formation of short-circuited turns in which the flux would set up a large circulating current. See Figure 9-5. The magnetic shield reduces the reluctance of the magnetic path, causing the normal current reactance to be approximately 15% to 20% higher than at the time of a short circuit. Sufficient iron is used for normal

MAGNETIC SHIELD CONSISTING OF LAMINATED SHEET STEEL

ABB Power T&D Company Inc.

Figure 9-5. A magnetic shield is placed around the outside of an oil-immersed current-limiting reactor coil to prevent the flux from flowing into the wall of the tank.

operation only. During a short circuit, the iron saturates and the reactor performs substantially the same as an air-core reactor. The increase in reactance at normal current flow is not objectionable because the rating of the reactor is based on the short-circuit reactance.

An oil-immersed reactor has a high factor of safety against flashover not obtained in any other way. This feature makes it desirable for outdoor service and for high-voltage applications. The high thermal capacity of the oil and the ease of cooling compared to the cooling of a dry-type reactor make it desirable for large units or locations where space is limited. The general appearance of an oil-immersed current-limiting reactor is similar to that of an oil-insulated transformer. Standard oil-insulated transformer cases are often used to hold oil-immersed current-limiting reactors.

Cores

A reactor may have a gapped-core or may be coreless. A gapped-core has a subdivided limb of core steel with air gaps inside the coil. See Figure 9-6. The gapped-core is more advantageous as the loss evaluation rate increases, particularly at higher system voltages. This is due to the higher energy density that can be achieved in a gapped-core reactor compared to a coreless reactor.

A coreless reactor does not have a subdivided limb of core steel with air gaps inside the coil. A shell form coreless reactor is similar in proportions and coil design to shell-type transformers.

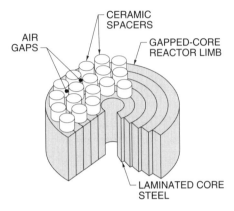

Figure 9-6. A gapped-core reactor has a subdivided limb of core steel with air gaps inside the coil.

The basic building block of a gapped-core is a cylindrical segment of radially laminated core steel, molded in epoxy resin to form a solid piece. These modules are stacked and cemented together with stiff ceramic spacers to make a solid core limb column. The magnetic field creates pulsating forces across the air gaps which amount to tens of tons. The reactor cores must be very stiff to eliminate objectionable vibrations. The radial lamination prevents flux from entering flat surfaces of core steel that would result in eddy-current overheating and hot spots.

Classification

Current-limiting reactors, being primarily protective devices, are connected as near as possible to the service or apparatus they protect. Current-limiting reactors are classified by use. Current-limiting reactors include feeder, paralleling, generator synchronizing, synchronizing, starting, DC smoothing, and bus reactors. See Figure 9-7.

- Feeder—Feeder reactors are used to localize voltage drop on the feeder where trouble occurs. They are used extensively because most short circuits occur in feeder circuits. Each feeder is connected to a bus through a reactor. Feeder reactors provide great protection to the system and make it possible to use small and inexpensive feeder circuit breakers and fewer bus supports. Feeder reactors also prevent the opening of the main breaker in case of trouble. This prevents cutting other feeders out of service due to a fault in one of the feeders.

- Paralleling—Paralleling reactors are current-limiting reactors used to correct the division of load between parallel-connected transformers that have unequal impedance voltages.

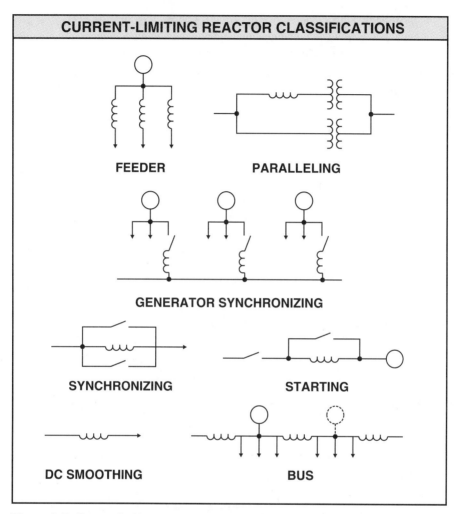

CURRENT-LIMITING REACTOR CLASSIFICATIONS

FEEDER PARALLELING

GENERATOR SYNCHRONIZING

SYNCHRONIZING STARTING

DC SMOOTHING BUS

Figure 9-7. Current-limiting reactors include feeder, paralleling, generator synchronizing, synchronizing, starting, DC smoothing, and bus reactors.

- Generator Synchronizing–Generator synchronizing reactors are connected in the leads between generators and station bus bars. When several large generators are paralleled at one station, it is often desirable to use generator reactors to prevent excessive damage to a winding due to the current rushes from the other machines in case of fault in the generator winding.

- Synchronizing–Synchronizing reactors are current-limiting reactors that are connected momentarily across the open contacts of a circuit-interrupting device for synchronizing purposes.

- Starting–Starting reactors are current-limiting reactors used to decrease the starting current of a machine or device.

- DC Smoothing–DC smoothing reactors are inductive reactors placed between the DC output of a thyristor converter and the load. A DC smoothing reactor limits the magnitude of ripple current in the load.

- Bus–Bus reactors are connected between different bus sections. See Figure 9-8. Generators are often tied together to a continuous, straight sectionalized bus to reduce the expense of generator reactors and also to eliminate the voltage drop under normal operating conditions. Each generator supplies its own group of feeder circuits. If another bus reactor is connected by the dotted line, it becomes a ring bus without generator reactors.

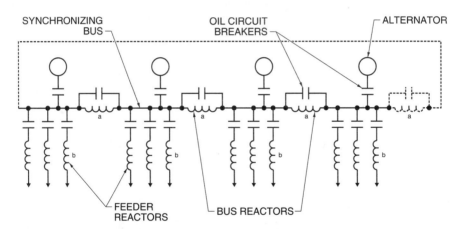

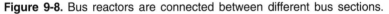

Figure 9-8. Bus reactors are connected between different bus sections.

A thorough study of present and future conditions of a particular system in which current-limiting reactors are to be used is necessary to determine the most economical amount of reactance to be used and to decide whether the reactance should be located in the feeders, busses, or generator leads. In the calculation of the possible short-circuit conditions, the inherent reactance of the generators must be taken into consideration because the reactance of the generator is normally sufficient to offset the total reactance of the various branches of the system.

SATURABLE-CORE REACTORS

The value of the inductive reactance in an AC circuit must often be controlled. This may be accomplished by using a saturable-core reactor. A *saturable-core reactor* is a magnetic-core reactor in which the reactance is controlled by changing the saturation of the core through variation of a superimposed flux. See Figure 9-9.

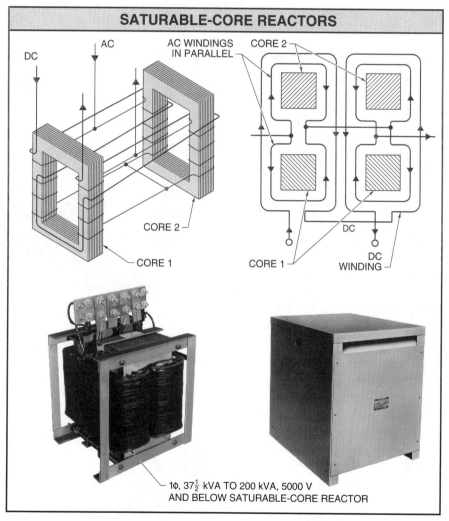

SATURABLE-CORE REACTORS

1φ, 37½ kVA TO 200 kVA, 5000 V
AND BELOW SATURABLE-CORE REACTOR

ABB Power T&D Company Inc.

Figure 9-9. A saturable-core reactor is controlled by changing the saturation of the core through variation of a superimposed flux.

The impedance of a saturable-core reactor varies cyclically with the AC. The performance of a saturable-core reactor is similar to a variable resistor in a DC circuit with the exception that the value of reactance (or impedance) is changed by controlling the magnetic saturation in an iron core. The reactance is controlled by circulating a DC in a separate winding. With no DC current flowing in the separate winding, the device functions as an iron core reactor. As the DC current is increased, the core becomes saturated with DC flux, thereby reducing the inductive reactance as far as AC is concerned. A complete range of control of the AC circuit is obtained in a smooth stepless manner. By using a saturable-core reactor, large amounts of AC power can be controlled by a small amount of DC power (normally 1% or less).

SHUNT REACTORS

A *shunt reactor* is a reactor that is connected in shunt to an electrical system. A shunt reactor is used to draw inductive current. See Figure 9-10. A shunt reactor is similar to a transformer. The normal use for shunt reactors is to compensate for capacitive currents from transmission lines, cables, or shunt capacitors.

ABB Power T&D Company Inc.

Figure 9-10. A shunt reactor is connected in shunt to an electrical system.

10 TRANSFORMER MAINTENANCE

Transformer safety standards include all basic electrical safety rules as well as additional rules specific to transformers. Costly transformer failures and interruption of service can be kept to a minimum if basic procedures are followed. An inspection checklist helps in the proper maintenance of transformers and regulators.

TRANSFORMER SAFETY STANDARDS

The basic safety rules that apply to all electrical equipment also apply to transformer installations with the exception that additional hazards are involved and additional safety precautions must be taken. See Figure 10-1. Always know and obey all posted regulations. The primary rule is to disconnect a transformer from all sources of electrical energy and ground the windings. Both the primary and the secondary must be disconnected to effectively disconnect a transformer. The secondary must be disconnected to eliminate any possibility of a secondary system feedback.

The disconnecting means must be locked in the open position after a transformer's primary and secondary windings have been disconnected. The fuses should be removed to an inaccessible area so they cannot be accidentally reinstalled if fuses or fuse cutouts are used. The primary and secondary windings should be grounded to discharge any capacitance energy that has been stored in the transformer when the transformer has been disconnected. This ground should not be removed until all work has been completed. The permanently installed transformer tank ground should be checked before starting work on the tank and cooling coils.

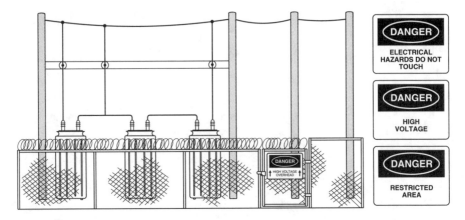

Figure 10-1. Safety precautions must be taken when working with transformers.

Special precautions must be taken to relieve the internal tank pressure before any attempt is made to remove the tank top or access cover if internal work or an internal inspection is necessary on liquid-immersed transformers. This is accomplished by partially operating a valve or plug above the liquid level. The transformer tank must be purged and a continuous source of clean air circulated in the tank if an inert gas is used.

Special precautions must be taken when working with transformers using askarel as an insulating liquid. Direct contact with askarel should be avoided because it has an irritating effect on the skin, especially the eyes, nose, and lips. Severe skin eruptions may develop by continuous absorption of askarel through the pores of the skin. Apply castor oil immediately to areas where continuous absorption of askarel is suspected.

Transformers containing askarel should not be opened when the transformer is hot because the fumes are toxic. If it is absolutely necessary to open a hot transformer, it should be done only in a well ventilated area. Direct exposure to the fumes should be avoided. When an individual enters a transformer tank, a second individual must be on duty outside the tank.

Transformers using PCB compounds as an insulating liquid need special attention by trained personnel. PCB materials used in transformers, oil circuit breakers, and other electrical equipment were declared hazardous in the mid-seventies and eliminated from the manufacture of new devices. PCBs may be found today in equipment manufactured before the mid-seventies and therefore, require handling by personnel trained to handle this type of material.

Special care must be taken to ensure that no tools or other equipment are dropped or left inside a transformer tank when work has been completed. A list of all tools and other articles used to make repairs should be listed before the job is started to ensure that an inventory can be taken when the job is completed. This ensures nothing was dropped or left inside the transformer tank.

Do not assume that a transformer is dead (de-energized). Ensure breakers, fuses, or switches are open before doing work. After work is completed, ensure all is clear before energizing.

Transformer Circuit Grounding

Secondary circuits of current transformers and potential transformers on circuits of primary voltage 460 V and above, shall be grounded in an approved manner.

Oil-Filled Equipment Handling

Precautions observed during oil-handling operations on oil-filled equipment, such as transformers and circuit breakers include:

- Equipment bonding–All apparatus tanks, shielded hoses, pumping or filtering equipment, drums, tank cars, trucks, and storage tanks shall be solidly bonded through a common interconnecting grounding cable.

- Grounding exposed conductors–All exposed conductors, such as transformer or circuit breaker bushings or coil ends of transformers which have bushings removed either physically or electrically, shall be connected to the same ground cable.

- Exceptions–For treatment of insulating oil in equipment under energized conditions, grounding is not required on exposed conductors but equipment bonding shall be made as specified.

- Oil-filtering jobs–When returning to a partially completed oil-filtering job after shutdown for any reason, all switching, grounding, and bonding shall be checked before resuming operations.

Current transformers shall not be worked on while energized, except under the immediate direction of an authorized person who shall remain at the job site until the job is completed. Current transformers must be shorted before any part of the circuit that they are installed in is opened.

Potential Transformer Clearances

The fuses or switches shall be opened on potential transformers or equipment to which potential transformers are connected when placing clearances. This prevents the cleared equipment from being energized from the low-voltage side.

Transformer Secondary Circuits

Secondary circuits of current transformers shall not be opened while the primary circuit is energized. The secondary terminals of the transformer must be effectively jumped by approved methods.

Entering Dangerous Areas or Vessels

Persons shall not enter a transformer or circuit breaker tank, oil storage tank or tank car, or other vessel which has contained flammable liquid until the vessel has been properly purged and ventilated. Prior to entry, tests for flammable or toxic vapors shall be made using approved indicators. Ventilation must be continued during the life of the job and a second person must be present before any one enters the vessel.

INSPECTION AND MAINTENANCE

Costly transformer failures and interruption of service can be kept to a minimum if basic procedures are followed. The most important factor is that the proper transformer is selected for each installation. When the proper transformer has been selected, the transformer must be properly installed.

Periodic inspection and maintenance is required to prevent costly failures and interruption of service after the transformer has been properly installed. It is often assumed that there is no need for inspection and maintenance because transformers have no moving parts. A transformer that has been operating satisfactorily for many years may suddenly fail without warning or any outward sign of distress because of neglect. This causes expensive repairs or replacement and loss of production.

The proper maintenance of transformers consists of:

- Ensuring current-carrying components are operating in a moisture-free insulating liquid or area.
- Keeping all parts clean and protected from rust, dirt, and corrosion.
- Testing the winding insulation and the insulating and cooling liquid.
- Lubricating all moving parts.
- Keeping tight all enclosures containing insulating and cooling liquids and weatherproof enclosures.
- Inspecting and testing the protective and indicating devices.
- Inspecting the transformer internally.
- Inspecting the auxiliary equipment, such as fans, coolers, lightning arresters, and grounds.

The frequency of inspection and maintenance of transformers and their associated equipment varies depending on the transformer size, type, and use. Normally, large power transformers that are costly to replace and normally serve electrical equipment that is critical for plant production, receive more frequent inspections and maintenance than small lighting and distribution transformers that serve less critical circuits. Also, large power transformers normally have more associated equipment, such as cooling fans, forced oil pumps, or protective alarms that must be properly maintained to ensure a trouble-free installation.

Personnel responsible for the proper and continuous operation of a transformer installation should develop an inspection and maintenance schedule. Once a schedule has been established, it should be closely followed and the readings and conditions observed should be recorded in a log or inspection sheet. A detailed record should also be kept, listing the information relative to the installation, showing the transformer specifications and characteristics, past history, repairs and tests made, spare parts, etc.

Load and Voltage

Frequent voltage readings should be taken because the current of a transformer determines the amount of heat generated by the transformer.

Hourly readings should be taken and recorded if they are not continuously recorded on instrument charts.

Liquid Level

The proper liquid level must be maintained to prevent a transformer from overheating. Transformer liquid level should be checked daily. If there is any loss from evaporation or leakage, it should be replaced immediately and repairs made to prevent further loss.

Temperature

The ambient temperature of the area where the transformer is installed should be observed hourly for large power or substation transformer installations. A less frequent observation may be made for small lighting and distribution transformers feeding less critical circuits. Temperature is an important factor in determining the amount of load a transformer can carry because the ambient temperature added to the transformer temperature rise determines the operating temperature. Normally, transformers are designed to operate continuously at full load with an average ambient temperature of 30°C (86°F) for cooling air with a maximum ambient temperature not to exceed 40°C (104°F). See Figure 10-2.

TEMPERATURE CONVERSIONS	
Convert °C to °F	**Convert °F to °C**
$°F = (1.8 \times °C) + 32$	$°C = \dfrac{(°F - 32)}{1.8}$
Example: Convert 30°C to °F $°F = (1.8 \times °C) + 32$ $°F = (1.8 \times 30) + 32$ $°F = 54 + 32$ $°F = 86°$	**Example:** Convert 86°F to °C $°C = \dfrac{(°F - 32)}{1.8}$ $°C = \dfrac{54}{1.8}$ $°C = 30°$

Figure 10-2. Ambient temperature may be expressed as °C or °F.

The liquid temperature of a transformer should be monitored hourly. The temperature is normally 10°C to 15°C less than the hottest-spot winding temperature. The hottest-spot winding temperature should also be monitored hourly because it determines the aging or rate of deterioration of the insulation of the transformer. Some experts have calculated that for each 8°C temperature rise above the rated temperature of the transformer, the aging rate for Class A insulation doubles. According to this rule, if a transformer having Class A insulation operates at a hottest-spot temperature of 8°C above its rated temperature, the life of the transformer is cut by half.

The effect that heat has on a transformer is critical. Hourly inspections should be made and the observations noted of the ambient temperature, liquid temperature, and the hottest-spot winding temperature. The load should be decreased or the cooling should be increased if the maximum temperatures are exceeded. Immediate steps must be taken before serious damage occurs within the transformer if a gradual or sudden increase in temperature is observed from previous readings under the same conditions of load and ambient temperatures.

Air in Dry-Type Transformers. Heat may also be a problem with dry-type transformers, therefore, hourly inspections should be performed. The ambient temperature of the surrounding air should be noted and the temperature of the ingoing and outgoing cooling air should be carefully monitored. An obstruction to the cooling air or a rise in the ambient temperature could result in serious damage to the transformer, especially if it is operating at rated load.

Water in Water-Cooled Transformers. Daily inspections should be performed to observe the temperature difference between the input and output of water through the cooling coils. The results should be monitored and recorded so that the efficiency of the cooling system can be determined. Semiannual checks should be performed to determine the water pressure and flow through the cooling coils. There may be an obstruction, such as scale (sediment) in the cooling coil if the water pressure remains constant but the flow decreases.

The water pumps, motors, and intake screens should be examined to ensure they are operating properly. The temperature of the input cooling water should not exceed 25°C, with the temperature difference between the input and output cooling water through the cooling coils not exceeding

10°C. The scale that has formed on the inside walls of the cooling system must be removed if the specified rate of flow cannot be achieved with the proper water pressure and all pumps operating properly. If proper heat transfer is not achieved after the cooling coils have been repaired and are functioning properly, the trouble is probably due to oil sludge deposits on the external surface of the cooling coils. The sludge must be removed for proper heat transfer. Special care should be taken so as not to expose the water in the cooling coils to a temperature that could cause the water to freeze, thereby causing severe damage to the transformer. If a water-cooled transformer is taken out of service and the transformer is exposed to below freezing temperatures, the water must be drained from the cooling coils.

Oil in Forced Oil-Cooled Transformers. The temperature of the oil entering and leaving the heat exchanger should be carefully observed and recorded daily. All other associated equipment, such as pumps and fans should also be monitored daily. A determination should be made that the cooling system is performing efficiently.

Pressure/Vacuum Gauge (Sealed Tank Transformers)

Most liquid-filled power transformers are equipped with a pressure/vacuum gauge that should be monitored daily. All readings should be recorded. The manufacturer's specified normal operating pressure range should be carefully followed. As the oil temperature changes, there should be a change in the pressure gauge reading. If the reading does not change, an air leak is normally indicated in the tank above the oil level.

Pressure Gauge (Gas-Oil-Sealed Preservation System)

The pressure gauge used with transformers that use an inert gas cushion above the oil level should be monitored daily and recorded. The manufacturer's specified operating range should be carefully followed. A leak in the system is indicated by a low-pressure reading when the transformer temperature is high, assuming the nitrogen cylinder pressure is sufficiently high. The nitrogen cylinder pressure gauge should also be monitored and recorded. The pressure of the nitrogen cylinder must not fall below the manufacturer's specifications.

A 200 cu ft cylinder normally lasts from six months to one year depending on the size and service conditions of the transformer. Caution should be taken to ensure that the oxygen content in the nitrogen cushion does not exceed 5% because a greater amount of oxygen creates an explosive mixture. The oxygen content of the mixture may be checked by an oxygen analyzer or indicator. The oxygen contents should be monitored weekly during the first month of operation of a new transformer and every six months thereafter.

Fans, Pumps, and Controls

All fans and pumps should be monitored daily for proper operation and the observation recorded. This includes the motors, fans, pumps, and associated control equipment. If a second stage fan is involved, but not frequently used, the motor, fan, and controls should be checked weekly. All auxiliary cooling units and their control equipment should be thoroughly cleaned and overhauled semiannually or more frequently if prevailing conditions warrant.

Dehydrating Breathers

The chemicals used in dehydrating breathers to remove moisture from the air entering a transformer should be monitored and the results recorded monthly. The two most commonly used chemicals are calcium chloride and silica gel. Silica gel breathers are replacing calcium chloride breathers. The advantage of using silica gel is that when silica gel becomes saturated with moisture, it does not restrict breathing as does calcium chloride. Silica gel, when dry, is blue in color. The color changes to pale pink as it becomes saturated with moisture. If the chemical is found to contain excessive moisture, it must be replaced or reconditioned. Silica gel can be dried by heating it in an open container at a temperature of between 150°C and 200°C for two hours. Calcium chloride can be dried by heating it in the same manner at a temperature of between 180°C and 200°C until completely dry. All air paths in breathers should be checked monthly to ensure they are free from any restrictions.

Pressure Relief Devices

A gas is generated when an arc occurs under the liquid inside a transformer. If this arc is severe, such as that caused by a short circuit, the amount of gas generated could create sufficient pressure to rupture the transformer tank. The rate of rise of gas pressure in a transformer is proportional to the arc power and inversely proportional to the volume of the gas space. Most liquid-filled transformers are equipped with pressure relief devices, such as pressure relief diaphragms or mechanical relief devices to relieve excessive pressure before serious damage is done to the transformer tank.

Pressure Relief Diaphragms

Pressure relief diaphragms are normally constructed of a thin layer of a suitable material, such as micarta, glass, bakelite, or thin metal. Pressure relief devices are normally mounted on the end of a large pipe connected to the transformer case above the liquid level. The diaphragm is normally designed to rupture from 10 psi to 15 psi. These devices should be monitored and the condition recorded quarterly.

When inspecting pressure relief diaphragms, close attention should be given to gasketed joints for evidence of rust, corrosion, and tightness. The condition of the diaphragms should be observed to determine if there are any defects or cracks. A defective gasket or a cracked diaphragm should be replaced immediately.

Mechanical Relief Devices

Mechanical relief devices are normally furnished as standard equipment with transformers using an inert-gas cushion oil-preservation system and are optional with liquid-filled transformers. Mechanical relief devices should be carefully monitored and the results recorded quarterly. Careful inspection should be made of the condition of the diaphragm and gasketed joints for evidence of leaks. If the device includes a visible or audible alarm, it should be carefully observed to ensure that it functions properly if the pressure in the transformer tank reaches a predetermined value.

Gas Detector and Pressure Relays

Gas detector relays are normally used with expansion tank (or conservator-type) transformers to detect developing faults or localized overheating of insulating materials. Gas detector relays are located in the pipe between the expansion tank and the transformer and are designed to trap any gas which may rise through the oil. The ability of this relay to measure a small accumulation of gas enables it to detect faults in their beginning stage. In normal practice, the accumulator contacts are connected to an alarm, but they may also be connected to trip the transformer off the line. These relays and their associated alarms should be monitored quarterly and their status recorded.

Pressure relays are designed to operate on a rapid rise in pressure in the transformer tank, normally caused by an arcing fault in the transformer. An equalizing orifice in the relay permits a slow change in pressure due to normal loading and ambient temperature without operating the alarm contacts. These relays and their alarms should be monitored quarterly and their status recorded.

Protective Alarms

A monthly inspection should be made and recorded of all transformer alarm and trip circuits. The settings and operation of the alarm actuating devices, such as pressure relays, pressure relief devices, oil flow indicators, liquid and hottest-spot winding temperature indicators, and outlet air thermometers should be monitored and recorded monthly.

Tap Changers

Most transformer installations include a no-load or load-type tap changer which permits changes in the voltage ratio of the transformer. Normally, a no-load tap changer is manually operated, but must not be operated while the transformer is energized. The load-type tap changer may be manually or automatically operated while the transformer is in operation. The original position of the tap changer should be carefully observed and the results recorded. When all work has been completed, the tap changer must be returned to its original position before the transformer is put back in service.

The exposed external mechanical parts should be carefully examined annually for wear, proper lubrication, corrosion, dirt, looseness, adjustment, and seal leakage. If there have been any indications of unsatisfactory operating conditions, a voltage ratio check should be conducted and an internal inspection made.

To conduct an internal inspection, the transformer must be opened and the oil level lowered to expose the electrical and mechanical parts of the tap-changer assembly. The contacts should be carefully examined and cleaned. The tap-changer operating mechanisms, such as motors, gears, shafts, bearings, and control relays should be carefully monitored and adjusted for proper operation. The tap changer should be rotated over its entire range several times and a careful inspection made to ensure that a complete contact surface and tension exists at each position. A further check may be made at each contact position by using a Wheatstone bridge to detect a poor contact. The manufacturer's specifications and instructions for the operation and maintenance of the equipment should be followed carefully.

Ground Connections

The noncurrent-carrying metal parts of a transformer installation are required by the NEC® to be effectively grounded. Conductive materials enclosing electric conductors or equipment, or forming part of such equipment are grounded for the purpose of preventing a voltage above the ground on these materials. Circuits and enclosures are grounded to facilitate overcurrent device operation in case of insulation failure or ground faults.

Ground connections should be inspected monthly and the results recorded. The inspection should include close examination for loose, corroded, or broken connections. The connections should be observed for any unusual heating at the connection. An annual resistance-to-ground inspection should be made and recorded.

Lightning Arresters

Lightning arresters should be regularly inspected and cleaned. A monthly inspection should be made and recorded. This inspection should include a visual observation of dirt and foreign deposits on the arrester or any undue mechanical strain from insufficient slack in the line.

An annual inspection should be made and recorded. This inspection should include a complete overhaul and cleaning of arresters. All connections should be clean and tight and all broken or damaged parts should be replaced.

Overcurrent Protection Devices

The prime factor when selecting the value of the overcurrent protection for transformers are the requirements of the NEC®. The basic provisions found in Article 450 of the 1996 NEC® require that each askarel- and oil-insulated transformer (a bank of three 1φ transformers operating as a 3φ unit is considered one transformer) shall be protected by an individual overcurrent device in the primary connection, rated or set at not more than 250% of the rated primary current of the transformer. An exception to this is a transformer having an overcurrent device in the secondary connection, rated or set at not more than 250% of the rated secondary current of the transformer. Another exception is a transformer equipped with a coordinated thermal overload protection by the manufacturer. This type of transformer is not required to have an individual overcurrent device in the primary connection, provided the primary feeder overcurrent device is rated or set to open at a current value not more than six times the rated current of the transformer for transformers having not more than 6% impedance, and not more than four times the rated current of the transformer for transformers having more than 6% but not more than 10% impedance.

The maximum overcurrent protection permitted by the NEC® does not protect a transformer from overheating resulting from a constant overload of up to 250% of the full-load current. These overcurrent devices protect the transformer installation from excessive currents caused by lightning, short circuits, or grounds. The transformer is protected from overheating by other protective devices, such as the hottest-spot winding indicator, liquid temperature indicator, or pressure relay.

The overcurrent devices selected normally consist of circuit breakers, fuse links, fuse cutouts, fused switches, oil circuit reclosers, or circuit breakers controlled by protective relays. The overcurrent protective devices and their associated equipment should be inspected monthly for evidence of loose connections, dust, dirt, corrosion, or heating. An annual inspection should be performed, at which time all equipment is carefully examined and cleaned. All manual and automatic switches and circuit breakers should be operated several times to ensure that all moving parts are free and operable.

Insulating Liquids

The proper inspection, testing, and maintenance of transformer insulating liquids is the most positive way to keep a transformer in proper operating condition and to extend the life expectancy of the transformer to a maximum value. Oxygen in the air in contact with insulating oil causes the oil to oxidize when heated, resulting in the formation of acids and sludge. The acid content of the insulating oil is a measure of its deterioration and its tendency to form sludge. The sludge is extremely harmful to a transformer because it settles on the internal parts of the transformer and greatly impedes cooling while reducing the flashover value of exposed parts and insulators. A sludge deposit of $\frac{1}{8}''$ to $\frac{1}{4}''$ on the core and coils may increase the temperature from 10°C to 15°C.

Moisture present in the insulating oil forms conducting paths on nonconducting parts, such as bushing shanks and insulated leads. This creates a definite hazard to the safe operation of the equipment. Moisture reduces the dielectric strength of all insulating liquids and may also be absorbed by the insulating materials used throughout the transformer. The dielectric strength of an insulating liquid (oil or askarel) is the voltage required to break down the liquid in a $0.1''$ gap between two electrodes of standard dimensions under specifically controlled testing conditions.

Askarel has the advantage as a cooling and insulating liquid in that it is neutral. Under normal operating conditions, it does not oxidize or decompose to produce acid or sludge. If, when askarel is tested, a small acidity content is found, this is not due to deterioration of the askarel, but is normally caused by its solvent action on certain organic insulating materials in the transformer.

Many plant electricians consider the dielectric test to be the most important. Other tests are available which are of considerable value. Because transformers need good oil for trouble-free operation, periodic tests of oil purity should be made and purification processes carried out promptly. The interval between oil tests depends on the load carried and the importance of the unit. Power station transformers and those serving vital circuits in industrial plants should have an oil test at three- to six-month intervals. Other transformers should receive oil tests annually. A properly equipped laboratory should be used for the accurate testing of oil. See Figure 10-3.

Kemper Insurance Co.

Figure 10-3. Inspection prevents costly damage caused by deteriorating oil.

In small, low-voltage transformers, operating companies periodically drain a small portion of the oil from the bottom of the tank and replace it with more oil at the top. This removes water which may have collected on the bottom. On large units, all the oil may be withdrawn and reconditioned to remove moisture and sludge held in suspension. This is done by pumping the oil through a filter press or by passing it through a centrifugal machine which separates the moisture and objectionable material from the oil.

Samples of the insulating liquid should be taken after the liquid has been given time to settle. The larger the transformer, the more time required. Normally, 8 hours is sufficient for a small transformer containing a barrel of liquid, while several days are necessary for a large transformer. If samples are taken from a transformer in operation, a sufficient amount should be taken to ensure that it actually represents the insulating liquid circulating through the transformer. Askarel samples should be taken from the top and bottom of the tank because askarels are heavier than water.

Insulating oil is not heavier than water, therefore the samples should be taken from the bottom of the tank. A sufficient amount should be withdrawn to ensure that it is truly a representative sample of the liquid in the transformer tank. A sufficient quantity of liquid must be drained for testing. For dielectric tests, at least 1 pint is required. For other tests, at least 1 quart is required.

Special precautions must be taken when taking and containing the samples. The samples should be taken on a clear, dry day and put into a clear glass or polyethylene container. A glass container enables the observation of water if it is present in the liquid. The container must be thoroughly cleaned with gasoline or Stoddard solvent and dried. The container should be further cleaned with soap and water and rinsed with distilled water and dried. If a tube or hose is required to remove samples from the bottom of a transformer tank, special care must be taken to use one constructed of glass or polyethylene. A tube or hose made of rubber should not be used because rubber normally contains sulphur which dissolves in the liquid and attacks the copper windings.

If the insulating oil or askarel tests are inadequate for safe operation, they should be reconditioned or replaced immediately. In all cases, the manufacturer's specifications should be followed.

Insulation Tests

If the insulating liquid in a transformer is kept free from moisture and the transformer has not been subjected to excessive voltages, overload, or excessive heating, the deterioration of the insulating materials within the transformer is kept to a minimum. The condition of the insulation between the separate windings and the windings and ground should be tested annually as part of preventive maintenance. This test is recommended even if the transformer has not been subjected to any of the factors that contribute to the deterioration of the insulating materials within the transformer. If the transformer has been exposed to an excessive voltage (such as lightning) or overload, has been overheated, or through normal inspection an excessive moisture content is found in the insulating liquid, immediate tests should be made to determine the condition of the transformer insulation.

Insulation Resistance. The insulation resistance values vary with the temperature and the amount of moisture in the insulation. The insulation resistance values vary inversely with the temperature. The individual who interprets the measurements should have considerable experience and judgment in the field. The insulation resistance values vary in proportion to the moisture content. The measurement is also affected by the type and size of the transformer being tested, the value of the test voltage, and the length of time the voltage is applied.

The temperature, humidity, and other similar factors should be recorded at the time the test is made because of the many variables involved when making these tests. These factors should be compared when making future tests to determine the trend of the insulating resistance. The tests are normally made with a megohmmeter. They measure the resistance of the insulation between the various windings of the transformer and between the various windings and ground.

Dielectric Absorption Test. To further test the condition of the insulation, a dielectric absorption test is performed. In a dielectric absorption test, a voltage is applied over a period of time (normally 10 minutes) to the equipment under test. At one minute intervals during the test, insulation resistance readings are taken and recorded on log-log graph paper with insulation resistance and time as the vertical and horizontal scales. The resultant curve on the graph of the insulation resistance with time gives an indication of the condition of the insulation under test. For proper testing, the equipment under test must be grounded before and between tests.

The insulation is in good condition if the dielectric absorption curve that has been plotted on the graph at one minute intervals is a straight line that increases with time. The insulation is unsafe and immediate repairs are required if the curve does not rise with time or only rises slightly and levels off.

The insulation resistance and dielectric absorption tests are only two of many tests that can be made to determine the condition of the insulation. There are many other tests that can be made but they are more complicated and normally require additional testing equipment. Unless the person performing the test and evaluating the test results has considerable experience, a qualified transformer consultant should be called when the tests indicate a dangerous condition before the transformer is taken out of service and expensive overhauling is started.

Detailed External Inspection. A semiannual inspection should be made of the entire transformer installation. A detailed list of all of the component parts should be made and each item that is inspected should be checked off to ensure maintenance personnel that no parts were overlooked. This inspection should be made to determine that all parts are free from damage, leakage, deterioration, or of accumulation of foreign deposits.

The operating area of a transformer installation normally dictates the amount and frequency of maintenance work that must be performed to ensure a trouble-free installation. Areas where carbon or metallic dusts are present require special precautions because these dusts are conductive and could cause a flashover. A troublesome area outdoors is where grain or other similar dusts are present. Although these dusts may not be conductive, a combination of the accumulation of these dusts and a high ambient moisture, such as fog, rain, or sleet could cause a flashover.

Bushings and potheads should be closely inspected to detect any chips, cracks, dirt, or dust accumulations. The porcelain may be cleaned by an oil-free solvent, such as an ammonia and water solution or carbon tetrachloride. This must be applied with a soft non-abrasive material such as a cloth. Because the fumes from carbon tetrachloride are toxic, it can only be used in a well-ventilated area. Solvents are available, such as a special silicone, that can be applied to porcelain bushings to retard the accumulation of dust and dirt. If the bushing contains an oil or compound, it should be examined for evidence of leakage. If leakage is detected, the equipment should be checked for moisture content and the manufacturer's specifications should be closely followed before adding compound or oil.

All metal parts of the transformer should be cleaned and examined for rust or corrosion. These parts should be kept painted. All gasket and pipe joints must be examined for leakage and kept in good repair. All electrical connections should be inspected for corrosion or evidence of heating. These connections should be checked for tightness because a poor connection in an electrical system under load causes heat and an eventual breakdown of the system.

Internal Inspection of Core and Coils

In open-type liquid-immersed power transformers that have a removable main cover or access cover, the cover should be removed and inspected annually. Any internal pressure must be relieved by cracking a valve or pressure relief device before removing the cover. The inside of the cover and the bushing supports should be closely examined for any evidence of condensation or rust that indicate a moisture problem. Immediate repairs must be made to eliminate this problem if there is evidence of moisture or rust.

Every five years, an internal inspection should be made of the top of the core and coil assembly of open-type liquid-immersed power transformers. This is done by dropping the liquid level to expose the top of the core. The core is then inspected for sludge deposits, loose connections, the condition of the insulation, or damaged or defective parts. All of the liquid must be removed and the necessary repairs made if there is any indication of trouble. The frequency of this type of inspection is determined by the operating conditions, age, and type of the transformer. This inspection may be performed at less frequent intervals if the transformer operates at normal loads, is of the inert-gas cushion type, and all of the regular maintenance checks indicate that the transformer is in good operating condition. This inspection should be performed at more frequent intervals if the transformer has been exposed to an overvoltage or overloads, or the regular maintenance tests indicate a problem.

An experienced transformer specialist should be consulted unless a person has the necessary experience and qualifications to make judgment decisions on questionable conditions of transformers. The manufacturer's specifications should be closely followed when making tests or repairs to transformers. Transformer maintenance tips include:

- Inspect a transformer thoroughly as soon as it is received from the manufacturer.

- Never energize a distribution transformer unless the oil is at the proper level.

- Handle transformers with care, avoiding severe jarring that might damage the internal structure.

- Inspect transformers before installation. Bushings, access plates, and other fittings should be inspected and thoroughly tightened before the transformer is placed in operation.

- Care must be taken so that moisture does not enter the transformer and nothing falls into the tank when replenishing oil.
- Never move or lift transformers by the bushings or other attachments. Lifting lugs are provided for this purpose.
- With pole-mounted transformers, care should be taken to see that the transformer is in a vertical position and securely fastened so that a severe impact against the pole does not jar it loose.
- Inspect all bushings, connections, protective devices, gaskets, the paint finish on the tank, and all other exterior fixtures for evidence of rust, corrosion, or overheating. The oil level should be checked, and the condition of the oil tested.
- Properly replace any inspection covers removed from a transformer for inspection. Replace any gaskets if not in perfect condition.
- Open the connections on the high and low voltage sides to avoid danger from feedback voltage when working with any transformer that is connected with any other transformer.

Transformer Failure

When a transformer fails, disconnect the transformer from all sources of power. Smoke, noise, or liquid may be coming from the tank which might give some indication as to what has failed, such as a winding, internal short, or that the transformer is overheating. Do not re-energize the transformer if it has been determined that a winding has failed. After disconnecting the source and load, make careful inspections and tests. When inspecting, look for external damage to the bushings, leads, and potheads. Check the level of liquids, take the temperature of liquids that can be measured, look for evidence of leakage, and look for evidence on the primary and secondary sides of the transformer for possible causes of failure. If possible, perform a dielectric strength test of the liquid from the top and bottom of the transformer.

INSPECTION CHECKLIST

An inspection checklist helps in the proper maintenance of transformers and regulators. See Figure 10-4. The procedures shown in Figure 10-4 are discussed on pages 282–286.

TRANSFORMER AND REGULATOR INSPECTION CHECKLIST

Item	Large Units Attended		Large Units Unattended		Small Units Attended		Small Units Unattended		Procedure Reference
Foundation, Rails, and Trucks		A		A		A		A	12.01
Tanks and Radiators	D	A	W	A	W	A	M	A	12.02
Oil and Water Piping	D	A	W	A	W	A	M	A	12.03
Valves and Plugs	D	A	W	A	W	A	M	A	12.04
Oil Level, Gauges, and Relays	D	A	W	A	W	A	M	A	12.05
Breathers and Vents	D	A	W	A	W	A	M	A	12.06
Relief Diaphragm	D	A	W	A	W	A	M	A	12.07
Water-Cooling Coils and Piping		A		A		A		A	12.08
Flow Indicators and Relays	D	A		A	D	A		A	12.09
Heat Exchangers		A							12.10
Oil Pumps	D	A							12.11
Cooling Fans	D	A			D	A			12.12
Temperature Indicators and Relays	D	A	W	A	W	A	M	A	12.13
Inert Gas Tanks	D		W						12.14
Gas Regulator, Gauges, and Relays	D	A	W	A					12.15
Gas Piping and Valves		A		A					12.16
Gas Analysis		Q		Q					12.17
Bushings	W	A	W	A	W	A	M	D	12.18
Bushing Current Transformers and Potential Device		A		A					12.19
Main Terminal and Ground Connections	D	A	W	A	W	A	M	A	12.20
Core and Coils		NS		NS		NS		NS	12.21
Internal Inspection		A		A		A		A	12.22
Terminal Board and Connections		A		A		A		A	12.23
Ratio Adjuster	W	A	W	A	W	A	M	A	12.24
Tap Changer or Regulator	D	A	W	A					12.25
Motor and Drive		A		A					12.26
Auxiliary and Limit Switches		A		A					12.27
Position Indicators		A		A					12.28
Operation Counter	W	A	W	A					12.29
Operation		A		A					12.30
Power Supplies and Wiring	D	A	W	A					12.31
Insulation Resistance		A		A		A		A	12.32
Oil Dielectric		A		A		A		A	12.33
Oil Acidity		5 yr		5 yr		5 yr		5 yr	12.34
Filter and Reclaim Oil		NS		NS		NS		NS	12.35
Fire Protection	M	A							12.36

D–routine daily inspection   M–routine monthly inspection   A–annual inspection
W–routine weekly inspection   Q–quarterly inspection   NS–not scheduled

United States Department of the Interior

Figure 10-4. An inspection checklist helps in the proper maintainence of transformers and regulators.

Foundations, Rails, and Trucks–12.01. *Annual.* Monitor the foundation for cracking and settling. A slight shift of the transformer may break bushings or connecting oil or water lines. Ensure that rail stops are firmly in place to hold transformer in position on the rails. Monitor transfer car and matching of the rails with transformer deck rails at each position. Paint metalwork as needed.

Tanks and Radiators–12.02. *Daily, weekly, or monthly.* Monitor for unusual noise and oil and water leaks.

Oil and Water Piping–12.03.

Valves and Plugs–12.04. *Annual.* Clean dirt and oil from radiating surfaces. Repaint as necessary. Stop excessive vibration of radiator tubes. Tighten loose or vibrating parts. Monitor for unusual internal noises. Inspect oil and water piping, valves, and plugs. Manipulate radiator cutoff valves to see that they are in operating condition and secure in the open position. Ensure that all oil drain valves which can be operated without wrenches are plugged or locked to prevent unauthorized opening.

Oil Levels, Gauges, and Relays–12.05. *Daily, weekly, or monthly.* Inspect oil level in main and auxiliary tanks, oil-filled bushings, etc. Changes in oil levels from time to time should be noted, taking into consideration the change in level caused by change in oil temperature. A rise in level in a water-cooled transformer (for a given temperature) indicates that water is leaking from the cooling coils into the oil.
Annual. Clean dirty gauge glasses and connections into tank. Monitor oil level indicators and relays for proper operation. Replenish oil if below normal. Drain and replace bushing oil if dirty or discolored.

Breathers and Vents–12.06. *Daily, weekly, or monthly.* Ensure that relief diaphragm has not opened and breather and vent appear normal.

Relief Diaphragm–12.07. *Annual.* Ensure that relief diaphragm is in operating condition and closes tightly. The nonshattering-type diaphragm should be actuated to see that it is not stuck shut from rust or paint. Ensure that the material used in the shattering-type diaphragm is not too thick or tough to be broken by reasonable internal pressure. Ensure that screens and baffles in vents or breathers are not obstructed or broken. If breathers are of the dehydrating type, monitor chemicals and replace if depleted.

Water Cooling Coils and Piping–12.08. *Annual.* Inspect external supply and drain piping for leaks. Flush out cooling coils or heat exchanger water passages with air and water. Test coils for leaks by applying air pressure to coils and observing for bubbles rising in oil and drop in air pressure with supply valve closed, or use a hydrostatic pressure test. A pressure of about 75 psi is recommended. If water scale is present, circulate a solution of 25% hydrochloric acid and water through the coils until clean. Then flush out thoroughly. Clean external surfaces of coils.

Flow Indicators and Relays–12.09. *Daily.* Ensure that proper supply of cooling water is flowing.
Annual. Check waterflow indicators and relays for proper operation.

Heat Exchangers–12.10. *Annual.* Clean and test water tubes similar to cooling coils. Inspect for oil and water leaks.

Oil Pumps–12.11. *Daily.* Ensure that oil-circulating pumps are in operation when required.
Annual. Inspect for proper operation and leakage.

Cooling Fans–12.12. *Daily.* Ensure that fans are in operation when necessary.
Annual. Inspect motors and controls.

Temperature Indicators and Relays–12.13. *Daily, weekly, or monthly.* Monitor and record transformer temperatures.
Annual. Inspect calibration of temperature indicators and relays. Inspect and clean relay contacts and operating mechanism.

Inert Gas Tanks–12.14. *Daily or weekly.* Monitor gas pressure left in tanks and change out at about 25 lb pressure.

Gas Regulator, Gauges, and Relays–12.15. *Daily or weekly.* Ensure that proper gauge pressure is being maintained in transformer.
Annual. Monitor setting and operation of regulator and relay. Ensure that gauges are indicating properly.

Gas Piping and Valves–12.16. *Annual.* Monitor for gas leaks by applying liquid soap on all joints, valves, connections, etc. with gas pressure raised to the maximum recommended by the transformer manufacturer.

Gas Analysis–12.17. *Annual.* Clean and test water tubes similar to cooling coils. Inspect for oil and water leaks.
Quarterly. Inspect analyzer for proper operation. Analyze gas. Purge if oxygen content is over 5%.

Bushings–12.18. *Weekly or monthly.* Inspect for chipped or broken porcelain, excessive dirt film, oil level, and oil or compound leaks.
Annual. Clean porcelain with water, chlorothene, or other suitable cleaner. Repair chipped spots by painting with lacquer such as red glyptal. Inspect gaskets for leaks. Tighten bolts. Monitor power factor. Inspect oil sample from bottom of bushing for dielectric strength and presence of water which may be entering at top. Replace or replenish oil if necessary.

Bushing Current Transformers and Potential Device–12.19. *Annual.* Inspect tap setting and adjustment at terminal board to see that they agree with diagrams. Monitor insulation resistance of wiring with devices connected. Monitor ratio and phase-angle adjustment of potential devices if changes have been made in secondary connections and burden. Tighten connections, including potential device tap, into bushing.

Main Terminal and Ground Connections–12.20. *Daily, weekly, or monthly.* Inspect for presence of foreign materials, birds' nests, etc. in or near connecting buswork, loose or heating connections, and loose or broken tank ground connections.
Annual. Tighten all bus and ground connections. Refinish joint contact surfaces if they have been overheating. Inspect ground cable to see that it is not loose or broken.

Core and Coils–12.21. *Not scheduled.* If the transformer has been properly maintained and has had no overheating or internal failure, it should not require untanking within its normal life. If sludge has been allowed to form, due to overheating and oxidation of the oil, the transformer should be untanked and the core, coils, oil passages, tank, and water-cooling coils washed down with clean oil under pressure to remove sludge and other accumulations which prevent proper circulation of the oil. Inflammable liquids should not be used in cleaning the core, coils, or inside of tank. Provide sufficient fresh air for workers while working inside of tank. While untanked, check for loose laminations, core bolts, insulating blocks, etc., and other pertinent features on the check list.

Internal Inspection–12.22. *Annual.* Lower the oil level to at least the top of the core. Inspect for sludge on core and windings. Inspect underside of cover for moisture and rust and then clean. Inspect connections at terminal board. Tighten all bolted connections, core bolts, etc. within reach.

Terminal Board and Connections–12.23.

Ratio Adjuster–12.24. *Weekly or monthly.* Note position of ratio adjuster and that it is adequately locked to prevent unauthorized operation. *Annual.* Inspect contacts and clean if reachable on internal inspection. If not reachable for visual inspection, monitor each position with Wheatstone bridge across winding to detect poor contact. Work adjuster back and forth over complete range several times.

Tap Changer or Regulator–12.25. *Daily or weekly.* Note position of tap changer. Ensure that positions on all three transformers are the same and that the manual operating device is locked. *Annual.* Drain oil from contact compartment. Clean and refinish contact surfaces. Inspect contact spring pressure. Inspect contact operating mechanism. Tighten connections and other bolts.

Motor and Drive–12.26. *Annual.* Inspect the motor. Inspect and adjust the brake. Inspect gears, shafts, and lubrication.

Auxiliary and Limit Switches–12.27. *Annual.* Monitor the condition of contacts and refinish if burned or corroded. Inspect contact springs, operating rods, and levers. Inspect closing and opening position with respect to position of main contact.

Position Indicators–12.28. *Annual.* Ensure that positions indicated correspond to position of main contact. Inspect remote electrical position indicators for correct operation, obstruction to movement of pointer, etc.

Operation Counter–12.29. *Weekly.* Monitor and record reading of operation counter.
Annual. Inspect operation counter or correct registration.

Operation–12.30. *Annual.* Run tap changer or regulator through several complete cycles by both control relay and manual control. Observe contacts and mechanism for proper operation.

Power Supplies and Wiring–12.31. *Daily or weekly.* Ensure that all power, control, and alarm supply circuit switches are closed and fuses in place so that circuits are completed.
Annual. Inspect fuses or circuit breakers on all power, control, and alarm supplies to auxiliary equipment and devices. Monitor and tighten wiring connections at all terminal points. Inspect wiring for open circuits, short circuits, and damaged insulation. Monitor insulation resistance of wiring with devices connected.

Insulation Resistance–12.32. *Annual.* Monitor the insulation resistance between each winding and between each winding and ground. Disconnect all external leads at the bushing terminals, except where the connecting leads can be suitably isolated at adjacent disconnecting switches for this test. A similar test using a capacitance bridge is recommended where such an instrument is available.

Oil Dielectric–12.33. *Annual.* Inspect the dielectric strength of the insulating oil in the main and auxiliary tanks and oil-filled bushings.

Oil Acidity–12.34. The acidity of the insulating oil in the main tank should be monitored at intervals of not more than 5 years. Transformers operating at high temperatures or showing signs of sludging or dark color of the oil should be inspected more frequently. Oil may be checked in the field with a Gerin test kit or samples sent to the Denver laboratory.

Filter and Reclaim Oil–12.35. *Not scheduled.* The necessity for filtering and/or reclaiming the insulating oil depends on the results obtained from the oil dielectric and oil acidity test. It may be more economical to replace the oil in small transformers rather than filter or reclaim it.

Fire Protection–12.36. *Monthly.* Where a water spray system is used, see that control valves and automatic devices are in operating condition and water supply is available.
Annually. With the transformer de-energized, turn ON the water spray system. Observe fog nozzles for proper coverage and spray.

APPENDIX

DECIMAL AND METRIC EQUIVALENTS OF COMMON FRACTIONS OF AN INCH

Fraction	Decimal	mm	Fraction	Decimal	mm
1/64	0.01562	0.397	33/64	0.51562	13.097
1/32	0.03125	0.794	17/32	0.53125	13.494
3/64	0.04688	1.191	35/64	0.54688	13.891
1/16	0.06250	1.588	9/16	0.56250	14.288
5/64	0.07812	1.984	37/64	0.57812	14.684
3/32	0.09375	2.381	19/32	0.59375	15.081
7/64	0.10938	2.778	39/64	0.60938	15.478
1/8	0.12500	3.175	5/8	0.62500	15.875
9/64	0.14062	3.572	41/64	0.64062	16.272
5/32	0.15525	3.969	21/32	0.65625	16.669
11/64	0.17188	4.366	43/64	0.67188	17.066
3/16	0.18750	4.763	11/16	0.68750	17.463
13/64	0.20312	5.159	45/64	0.70312	17.859
7/32	0.21875	5.556	23/32	0.71875	18.256
15/64	0.23438	5.953	47/64	0.73438	18.653
1/4	0.25000	6.350	3/4	0.75000	19.050
17/64	0.26562	6.747	49/64	0.76562	19.447
9/32	0.28125	7.144	25/32	0.78125	19.844
19/64	0.29688	7.541	51/64	0.79688	20.241
5/16	0.31250	7.938	13/16	0.81250	20.638
21/64	0.32812	8.334	53/64	0.82812	21.034
11/32	0.34375	8.731	27/32	0.84375	21.431
23/64	0.35938	9.128	55/64	0.85938	21.828
3/8	0.37500	9.525	7/8	0.87500	22.225
25/64	0.39062	9.922	57/64	0.89062	22.622
13/32	0.40625	10.319	29/32	0.90625	23.019
27/64	0.42188	10.716	59/64	0.92188	23.416
7/16	0.43750	11.113	15/16	0.93750	23.813
29/64	0.45312	11.509	61/64	0.95312	24.209
15/32	0.46875	11.906	31/32	0.96875	24.606
31/64	0.48438	12.303	63/64	0.98438	25.003
1/2	0.50000	12.700	1/1	1.00000	25.400

MILLIMETER AND DECIMAL INCH EQUIVALENTS*

mm	in.	mm.	in.	mm	in.	mm	in.	mm	in.	mm	in.
1/50	.00079	26/50	.02047	1	.03937	26	1.02362	51	2.00787	76	2.99212
2/50	.00157	27/50	.02126	2	.07874	27	1.06299	52	2.04724	77	3.03149
3/50	.00236	28/50	.02205	3	.11811	28	1.10236	53	2.08661	78	3.07086
4/50	.00315	29/50	.02283	4	.15748	29	1.14173	54	2.12598	79	3.11023
5/50	.00394	30/50	.02362	5	.19685	30	1.18110	55	2.16535	80	3.14960
6/50	.00472	31/50	.02441	6	.23622	31	1.22047	56	2.20472	81	3.18897
7/50	.00551	32/50	.02520	7	.27559	32	1.25984	57	2.24409	82	3.22834
8/50	.00630	33/50	.02598	8	.31496	33	1.29921	58	2.28346	83	3.26771
9/50	.00709	34/50	.02677	9	.35433	34	1.33858	59	2.32283	84	3.30708
10/50	.00787	35/50	.02756	10	.39370	35	1.37795	60	2.36220	85	3.34645
11/50	.00866	36/50	.02835	11	.43307	36	1.41732	61	2.40157	86	3.38582
12/50	.00945	37/50	.02913	12	.47244	37	1.45669	62	2.44094	87	3.42519
13/50	.01024	38/50	.02992	13	.51181	38	1.49606	63	2.48031	88	3.46456
14/50	.01102	39/50	.03071	14	.55118	39	1.53543	64	2.51968	89	3.50393
15/50	.01181	40/50	.03150	15	.59055	40	1.57480	65	2.55905	90	3.54330
16/50	.01260	41/50	.03228	16	.62992	41	1.61417	66	2.59842	91	3.58267
17/50	.01339	42/50	.03307	17	.66929	42	1.65354	67	2.63779	92	3.62204
18/50	.01417	43/50	.03386	18	.70866	43	1.69291	68	2.67716	93	3.66141
19/50	.01496	44/50	.03465	19	.74803	44	1.73228	69	2.71653	94	3.70078
20/50	.01575	45/50	.03543	20	.78740	45	1.77165	70	2.75590	95	3.74015
21/50	.01654	46/50	.03622	21	.82677	46	1.81102	71	2.79527	96	3.77952
22/50	.01732	47/50	.03701	22	.86614	47	1.85039	72	2.83464	97	3.81889
23/50	.01811	48/50	.03780	23	.90551	48	1.88976	73	2.87401	98	3.85826
24/50	.01890	49/50	.03858	24	.94488	49	1.92913	74	2.91338	99	3.89763
25/50	.01969			25	.98425	50	1.96850	75	2.95275	100	3.93700

*Based on 1/100 mm = .003973″   10 mm = 1 centimeter = 0.3937″   25.4 mm = 1″
10 cm = 1 decimeter = 3.937″   10 dm = 1 meter = 39.37″

GREEK SYMBOLS

Alpha	A α	Nu	N ν
Beta	B β	Xi	Ξ ξ
Gamma	Γ γ	Omicron	O o
Delta	Δ δ	Pi	Π π
Epsilon	E ε	Rho	P ρ
Zeta	Z ζ	Sigma	Σ σ
Eta	H η	Tau	T τ
Theta	Θ θ	Upsilon	Y υ
Iota	I ι	Phi	Φ φ
Kappa	K κ	Chi	X χ
Lambda	Λ λ	Psi	Ψ ψ
Mu	M μ	Omega	Ω ω

TEMPERATURE CONVERSIONS

Convert °C to °F	Convert °F to °C
$°F = (1.8 \times °C) + 32$	$°C = \dfrac{(°F - 32)}{1.8}$

ENGLISH SYSTEM

		Unit	Abbr	Equivalents
LENGTH		mile	mi	5280', 320 rd, 1760 yd
		rod	rd	5.50 yd, 16.5'
		yard	yd	3', 36"
		foot	ft *or* '	12", .333 yd
		inch	in. *or* "	.083', .028 yd
AREA		square mile	sq mi *or* mi^2	640 A, 102,400 sq rd
		acre	A	4840 sq yd, 43,560 sq ft
		square rod	sq rd *or* rd^2	30.25 sq yd, .00625 A
		square yard	sq yd *or* yd^2	1296 sq in., 9 sq ft
		square foot	sq ft *or* ft^2	144 sq in., .111 sq yd
		square inch	sq in. *or* in^2	.0069 sq ft, .00077 sq yd
VOLUME		cubic yard	cu yd *or* yd^3	27 cu ft, 46,656 cu in.
		cubic foot	cu ft *or* ft^3	1728 cu in., .0370 cu yd
		cubic inch	cu in. *or* in^3	.00058 cu ft, .000021 cu yd
CAPACITY	*U.S. liquid measure*	gallon	gal.	4 qt (231 cu in.)
		quart	qt	2 pt (57.75 cu in.)
		pint	pt	4 gi (28.875 cu in.)
		gill	gi	4 fl oz (7.219 cu in.)
		fluidounce	fl oz	8 fl dr (1.805 cu in.)
		fluidram	fl dr	60 min (.226 cu in.)
		minim	min	$\frac{1}{6}$ fl dr (.003760 cu in.)
	U.S. dry measure	bushel	bu	4 pk (2150.42 cu in.)
		peck	pk	8 qt (537.605 cu in.)
		quart	qt	2 pt (67.201 cu in.)
		pint	pt	$\frac{1}{2}$ qt (33.600 cu in.)
	British imperial liquid and dry measure	bushel	bu	4 pk (2219.36 cu in.)
		peck	pk	2 gal. (554.84 cu in.)
		gallon	gal.	4 qt (277.420 cu in.)
		quart	qt	2 pt (69.355 cu in.)
		pint	pt	4 gi (34.678 cu in.)
		gill	gi	5 fl oz (8.669 cu in.)
		fluidounce	fl oz	8 fl dr (1.7339 cu in.)
		fluidram	fl dr	60 min (.216734 cu in.)
		minim	min	$\frac{1}{60}$ fl dr (.003612 cu in.)
MASS AND WEIGHT	*avoirdupois*	ton		2000 lb
		short ton	t	2000 lb
		long ton		2240 lb
		pound	lb *or* #	16 oz, 7000 gr
		ounce	oz	16 dr, 437.5 gr
		dram	dr	27.344 gr, .0625 oz
		grain	gr	.037 dr, .002286 oz
	troy	pound	lb	12 oz, 240 dwt, 5760 gr
		ounce	oz	20 dwt, 480 gr
		pennyweight	dwt *or* pwt	24 gr, .05 oz
		grain	gr	.042 dwt, .002083 oz
	apothecaries'	pound	lb ap	12 oz, 5760 gr
		ounce	oz ap	8 dr ap, 480 gr
		dram	dr ap	3 s ap, 60 gr
		scruple	s ap	20 gr, .333 dr ap
		grain	gr	.05 s, .002083 oz, .0166 dr ap

METRIC SYSTEM

	Unit	Abbr	Number of Base Units
LENGTH	kilometer	km	1000
	hectometer	hm	100
	dekameter	dam	10
	*meter	m	1
	decimeter	dm	.1
	centimeter	cm	.01
	millimeter	mm	.001
AREA	square kilometer	sq km *or* km^2	1,000,000
	hectare	ha	10,000
	are	a	100
	square centimeter	sq cm *or* cm^2	.0001
VOLUME	cubic centimeter	cu cm, cm^3, *or* cc	.000001
	cubic decimeter	dm^3	.001
	*cubic meter	m^3	1
CAPACITY	kiloliter	kl	1000
	hectoliter	hl	100
	dekaliter	dal	10
	*liter	l	1
	cubic decimeter	dm^3	1
	deciliter	dl	.10
	centiliter	cl	.01
	milliliter	ml	.001
MASS AND WEIGHT	metric ton	t	1,000,000
	kilogram	kg	1000
	hectogram	hg	100
	dekagram	dag	10
	*gram	g	1
	decigram	dg	.10
	centigram	cg	.01
	milligram	mg	.001

*Base Units

ENGLISH TO METRIC EQUIVALENTS

		Unit	Metric Equivalent
LENGTH		mile	1.609 km
		rod	5.029 m
		yard	.9144 m
		foot	30.48 cm
		inch	2.54 cm
AREA		square mile	2.590 k^2
		acre	.405 hectacre, 4047 m^2
		square rod	25.293 m^2
		square yard	.836 m^2
		square foot	.093 m^2
		square inch	6.452 cm^2
VOLUME		cubic yard	.765 m^3
		cubic foot	.028 m^3
		cubic inch	16.387 cm^3
CAPACITY	*U.S. liquid measure*	gallon	3.785 l
		quart	.946 l
		pint	.473 l
		gill	118.294 ml
		fluidounce	29.573 ml
		fluidram	3.697 ml
		minim	.061610 ml
	U.S. dry measure	bushel	35.239 l
		peck	8.810 l
		quart	1.101 l
		pint	.551 l
	British imperial liquid and dry measure	bushel	.036 m^3
		peck	.0091 m^3
		gallon	4.546 l
		quart	1.136 l
		pint	568.26 cm^3
		gill	142.066 cm^3
		fluidounce	28.412 cm^3
		fluidram	3.5516 cm^3
		minim	.059194 cm^3
MASS AND WEIGHT	*avoirdupois*	short ton	.907 t
		long ton	1.016 t
		pound	.454 kg
		ounce	28.350 g
		dram	1.772 g
		grain	.0648 g
	troy	pound	.373 kg
		ounce	31.103 g
		pennyweight	1.555 g
		grain	.0648 g
	apothecaries'	pound	.373 kg
		ounce	31.103 g
		dram	3.888 g
		scruple	1.296 g
		grain	.0648 g

METRIC TO ENGLISH EQUIVALENTS			
	Unit	**British Equivalent**	
LENGTH	kilometer	.62 mi	
	hectometer	109.36 yd	
	dekameter	32.81′	
	meter	39.37″	
	decimeter	3.94″	
	centimeter	.39″	
	millimeter	.039″	
AREA	square kilometer	.3861 sq mi	
	hectacre	2.47 A	
	are	119.60 sq yd	
	square centimeter	.155 sq in.	
VOLUME	cubic centimeter	.061 cu in.	
	cubic decimeter	61.023 cu in.	
	cubic meter	1.307 cu yd	
CAPACITY		*cubic*	*dry*	*liquid*
	kiloliter	1.31 cu yd		
	hectoliter	3.53 cu ft	2.84 bu	
	dekaliter	.35 cu ft	1.14 pk	2.64 gal.
	liter	61.02 cu in.	.908 qt	1.057 qt
	cubic decimeter	61.02 cu in.	.908 qt	1.057 qt
	deciliter	6.1 cu in.	.18 pt	21 pt
	centiliter	.61 cu in.		338 fl oz
	milliliter	.061 cu in.		.27 fl dr
MASS AND WEIGHT	metric ton	1.102 t	
	kilogram	2.2046 lb	
	hectogram	3.527 oz	
	dekagram	.353 oz	
	gram	.035 oz	
	decigram	1.543 gr	
	centigram	.154 gr	
	milligram	.015 gr	

TRIGONOMETRIC FUNCTIONS

Angle	SIN	COS	TAN	COT	SEC	CSC	
0°	0.0000	1.0000	0.0000	∞	1.0000	∞	90°
1	0.0175	0.9998	0.0175	57.2900	1.0002	57.2987	89
2	0.0349	0.9994	0.0349	28.6363	1.0006	28.6537	88
3	0.0523	0.9986	0.0524	19.0811	1.0014	19.1073	87
4	0.0698	0.9976	0.0699	14.3007	1.0024	14.3356	86
5°	0.0872	0.9962	0.0875	11.4301	1.0038	11.4737	85°
6	0.1045	0.9945	0.1051	9.5144	1.0055	9.5668	84
7	0.1219	0.9925	0.1228	8.1443	1.0075	8.2055	83
8	0.1392	0.9903	0.1405	7.1154	1.0098	7.1853	82
9	0.1564	0.9877	0.1584	6.3138	1.0125	6.3925	81
10°	0.1736	0.9848	0.1763	5.6713	1.0154	5.7588	80°
11	0.1908	0.9816	0.1944	5.1446	1.0187	5.2408	79
12	0.2079	0.9781	0.2126	4.7046	1.0223	4.8097	78
13	0.2250	0.9744	0.2309	4.3315	1.0263	4.4454	77
14	0.2419	0.9703	0.2493	4.0108	1.0306	4.1336	76
15°	0.2588	0.9659	0.2679	3.7321	1.0353	3.8637	75°
16	0.2756	0.9613	0.2867	3.4874	1.0403	3.6280	74
17	0.2924	0.9563	0.3057	3.2709	1.0457	3.4203	73
18	0.3090	0.9511	0.3249	3.0777	1.0515	3.2361	72
19	0.3256	0.9455	0.3443	2.9042	1.0576	3.0716	71
20°	0.3420	0.9397	0.3640	2.7475	1.0642	2.9238	70°
21	0.3584	0.9336	0.3839	2.6051	1.0711	2.7904	69
22	0.3746	0.9272	0.4040	2.4751	1.0785	2.6695	68
23	0.3907	0.9205	0.4245	2.3559	1.0864	2.5593	67
24	0.4067	0.9135	0.4452	2.2460	1.0946	2.4586	66
25°	0.4226	0.9063	0.4663	2.1445	1.1034	2.3662	65°
26	0.4384	0.8988	0.4877	2.0503	1.1126	2.2812	64
27	0.4540	0.8910	0.5095	1.9626	1.1223	2.2027	63
28	0.4695	0.8829	0.5317	1.8807	1.1326	2.1301	62
29	0.4848	0.8746	0.5543	1.8040	1.1434	2.0627	61
30°	0.5000	0.8660	0.5774	1.7321	1.1547	2.0000	60°
31	0.5150	0.8572	0.6009	1.6643	1.1666	1.9416	59
32	0.5299	0.8480	0.6249	1.6003	1.1792	1.8871	58
33	0.5446	0.8387	0.6494	1.5399	1.1924	1.8361	57
34	0.5592	0.8290	0.6745	1.4826	1.2062	1.7883	56
35°	0.5736	0.8192	0.7002	1.4281	1.2208	1.7434	55°
36	0.5878	0.8090	0.7265	1.3764	1.2361	1.7013	54
37	0.6018	0.7986	0.7536	1.3270	1.2521	1.6616	53
38	0.6157	0.7880	0.7813	1.2799	1.2690	1.6243	52
39	0.6293	0.7771	0.8098	1.2349	1.2868	1.5890	51
40°	0.6428	0.7660	0.8391	1.1918	1.3054	1.5557	50°
41	0.6561	0.7547	0.8693	1.1504	1.3250	1.5243	49
42	0.6691	0.7431	0.9004	1.1106	1.3456	1.4945	48
43	0.6820	0.7314	0.9325	1.0724	1.3673	1.4663	47
44	0.6947	0.7193	0.9657	1.0355	1.3902	1.4396	46
45°	0.7071	0.7071	1.0000	1.0000	1.4142	1.4142	45°
	COS	**SIN**	**COT**	**TAN**	**CSC**	**SEC**	**ANGLE**

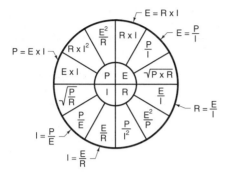

VALUES IN INNER CIRCLE
ARE EQUAL TO VALUES
IN CORRESPONDING
OUTER CIRCLE

OHM'S LAW AND POWER FORMULA

POWER FORMULA ABBREVIATIONS AND SYMBOLS	
P = Watts	V = Volts
I = Amps	VA = Volt Amps
A = Amps	ϕ = Phase
R = Ohms	$\sqrt{}$ = Square Root
E = Volts	

UNITS OF POWER

Power	W	ft lb/s	HP	kW
Watt	1	0.7376	1.341×10^{-3}	0.001
Foot-pound/sec	1.356	1	1.818×10^{-3}	1.356×10^{-3}
Horsepower	745.7	550	1	0.7457
Kilowatt	1000	736.6	1.341	1

POWER FORMULAS – 1ϕ, 3ϕ

Phase	To Find	Use Formula	Example Given	Example Find	Example Solution
1ϕ	I	$I = \dfrac{VA}{V}$	32,000 VA, 240 V	I	$I = \dfrac{VA}{V}$ $I = \dfrac{32{,}000 \text{ VA}}{240 \text{ V}}$ $I = \textbf{133 A}$
1ϕ	VA	$VA = I \times V$	100 A, 240 V	VA	$VA = I \times V$ $VA = 100 \text{ A} \times 240 \text{ V}$ $VA = \textbf{24,000 VA}$
1ϕ	V	$V = \dfrac{VA}{I}$	42,000 VA, 350 A	V	$V = \dfrac{VA}{I}$ $V = \dfrac{42{,}000 \text{ VA}}{350 \text{ A}}$ $V = \textbf{120 V}$
3ϕ	I	$I = \dfrac{VA}{V \times \sqrt{3}}$	72,000 VA, 208 V	I	$I = \dfrac{VA}{V \times \sqrt{3}}$ $I = \dfrac{72{,}000 \text{ VA}}{360 \text{ V}}$ $I = \textbf{200 A}$
3ϕ	VA	$VA = I \times V \times \sqrt{3}$	2 A, 240 V	VA	$VA = I \times V \times \sqrt{3}$ $VA = 2 \times 416$ $VA = \textbf{832 VA}$

AC/DC FORMULAS

To Find	DC	AC		
		1ϕ, 115 V or 220 V	1ϕ, 208 V, 230 V, or 240 V	3ϕ – All Voltages
I, HP known	$\dfrac{HP \times 746}{E \times Eff}$	$\dfrac{HP \times 746}{E \times Eff \times PF}$	$\dfrac{HP \times 746}{E \times Eff \times PF}$	$\dfrac{HP \times 746}{1.73 \times E \times Eff \times PF}$
I, kW known	$\dfrac{kW \times 1000}{E}$	$\dfrac{kW \times 1000}{E \times PF}$	$\dfrac{kW \times 1000}{E \times PF}$	$\dfrac{kW \times 1000}{1.73 \times E \times PF}$
I, kVA known		$\dfrac{kVA \times 1000}{E}$	$\dfrac{kVA \times 1000}{E}$	$\dfrac{kVA \times 1000}{1.763 \times E}$
KW	$\dfrac{I \times E}{1000}$	$\dfrac{I \times E \times PF}{1000}$	$\dfrac{I \times E \times PF}{1000}$	$\dfrac{I \times E \times 1.73 \times PF}{1000}$
kVA		$\dfrac{I \times E}{1000}$	$\dfrac{I \times E}{1000}$	$\dfrac{I \times E \times 1.73}{1000}$
HP (output)	$\dfrac{I \times E \times Eff}{746}$	$\dfrac{I \times E \times Eff \times PF}{746}$	$\dfrac{I \times E \times Eff \times PF}{746}$	$\dfrac{I \times E \times 1.73 \times Eff \times PF}{746}$

Eff = efficiency

HORSEPOWER FORMULAS

To Find	Use Formula	Example		
		Given	Find	Solution
HP	$HP = \dfrac{I \times E \times Eff}{746}$	240 V, 20 A, 85% Eff	HP	$HP = \dfrac{I \times E \times Eff}{746}$ $HP = \dfrac{240 \text{ V} \times 20 \text{ A} \times 85\%}{746}$ $HP = \mathbf{5.5}$
I	$I = \dfrac{HP \times 746}{E \times Eff \times PF}$	10 HP, 240 V, 90% Eff, 88% PF	I	$I = \dfrac{HP \times 746}{E \times Eff \times PF}$ $I = \dfrac{10 \text{ HP} \times 746}{240 \text{ V} \times 90\% \times 88\%}$ $I = \mathbf{39\ A}$

VOLTAGE DROP FORMULAS – 1ϕ, 3ϕ

Phase	To Find	Use Formula	Example		
			Given	Find	Solution
1ϕ	VD	$VD = \dfrac{2 \times R \times L \times I}{CM}$	240 V, 40 A, 60′ L, 16,510 CM, 12 R	HP	$VD = \dfrac{2 \times R \times L \times I}{CM}$ $VD = \dfrac{2 \times 12 \times 40 \times 60}{16,510}$ $VD = \mathbf{3.5}$
3ϕ	VD	$VD = \dfrac{2 \times R \times L \times I}{CM}$	208 V, 110 A, 75′ L, 66,360 CM, 12 R, .866 multiplier	I	$VD = \dfrac{2 \times R \times L \times I}{CM}$ $VD = \dfrac{2 \times 12 \times 75 \times 110}{66,360}$ $VD = \mathbf{2.58}$

$* \dfrac{\sqrt{3}}{2} = .866$

FULL-LOAD CURRENTS – 1φ TRANSFORMERS

kVA Rating	Voltage					
	120	208	240	480	600	2400
1	8.34	4.8	4.16	2.08	1.67	.42
3	25	14.4	12.5	6.25	5	1.25
5	41.7	24	20.8	10.4	8.35	2.08
7.5	62.5	36.1	31.2	15.6	12.5	3.12
10	83.4	48	41.6	20.8	16.7	4.16
15	125	72	62.5	31.2	25	6.25
25	208	120	104	52	41.7	10.4
37.5	312	180	156	78	62.5	15.6
50	417	240	208	104	83.5	20.8
75	625	361	312	156	125	31.2
100	834	480	416	208	167	41.6
125	1042	600	520	260	208	52
167.5	1396	805	698	349	279	70
200	1666	960	833	416	333	83.3
250	2080	1200	1040	520	417	104
333	2776	1600	1388	694	555	139
500	4170	2400	2080	1040	835	208

For other kVA ratings or voltages:

$$I = \frac{kVA \times 1000}{V}$$

FULL-LOAD CURRENTS – 3φ TRANSFORMERS

kVA Rating	Voltage (Line-to-Line)					
	208	240	480	600	2400	4160
3	8.3	7.2	3.6	2.9	.72	.415
6	16.6	14.4	7.2	5.8	1.44	.83
9	25	21.6	10.8	8.7	2.16	1.25
15	41.6	36	18	14.4	3.6	2.1
30	83	72	36	29	7.2	4.15
45	125	108	54	43	10.8	6.25
75	208	180	90	72	72	10.4
100	278	241	120	96	24	13.9
150	416	350	180	144	36	20.8
225	625	542	271	217	54	31.2
300	830	720	360	290	72	41.5
500	1390	1200	600	480	120	69.4
750	2080	1800	900	720	180	104
1000	2775	2400	1200	960	240	139
1500	4150	3600	1800	1440	360	208
2000	5550	4800	2400	1930	480	277
2500	6950	6000	3000	2400	600	346
5000	13,900	12,000	6000	4800	1200	694
7500	20,800	18,000	9000	7200	1800	1040
10,000	27,750	24,000	12,000	9600	2400	1386

For other kVA ratings or voltages:

$$I = \frac{kVA \times 1000}{V \times \sqrt{3}}$$

IEEE DEVICE NUMBERS
Switchgear and Pipeline Apparatus

No.	Device	No.	Device
1.	Master Element	48.	Incomplete Sequence Relay
2.	Time-Delay Starting or Closing Relay	49.	AC Thermal Relay
3.	Checking or Interlocking Relay	50.	Instantaneous Overcurrent or Rate-of-Rise Relay
4.	Master Contactor	51.	AC Time Overcurrent Relay
5.	Stopping Device	52.	AC Circuit Breaker
6.	Starting Circuit Breaker	53.	Exciter of DC Generator Relay
7.	Anode Circuit Breaker	54.	High-Speed DC Circuit Breaker
8.	Control Power Disconnect	55.	Power Factor Relay
9.	Reversing Device	56.	Field Application Relay
10.	Unit Sequence Switch	57.	Short-Circuiting or Grounding Device
11.	Control Power Transformer	58.	Power Rectifier Misfire Relay
12.	Over-Speed Device	59.	Overvoltage Relay
13.	Synchronous Speed Device	60.	Voltage Balance Relay
14.	Under-Speed Device	61.	Current Balance Relay
15.	Speed- or Frequency-Matching Device	62.	Time-Delay Stopping or Opening Relay
16.	Battery-Charging Device	63.	Liquid or Gas Pressure, Level or Flow Relay
17.	Shunting or Discharge Switch	64.	Ground Protective Relay
18.	Accelerating or Decelerating Device	65.	Governor
19.	Starting-to-Running Transition Contactor	66.	Notching or Jogging Device
20.	Electrically-Operated Valve	67.	AC Directional Overcurrent Relay
21.	Distance Relay	68.	DC Thermal Relay
22.	Equalizer Circuit Breaker or Contactor	69.	Permissive Control Device
23.	Temperature Control Device	70.	Electrically-Operated Rheostat
24.	Bus-Tie Breaker	71.	Emergency DC Line Circuit Breaker or Contactor
25.	Synchronizing Device	72.	DC Circuit Breaker
26.	Apparatus Thermal Device	73.	Load-Resistor Contactor
27.	AC Undervoltage Device	74.	Alarm Relay
28.	Resistor Thermal Device	75.	Position Changing Mechanism
29.	Isolating Contactor	76.	DC Overcurrent Relay
30.	Annunciator Relay	77.	Pulse Transmitter
31.	Separate Excitation Device	78.	Phase Angle Measuring or Out-of-Step Relay
32.	Direction Power Relay	79.	AC Reclosing Relay
33.	Position Switch	80.	DC Undervoltage Relay
34.	Motor-Operated Sequence Switch	81.	Frequency Relay
35.	Brush-Operating Slip Ring-Short Circuiting Device	82.	DC Reclosing Relay
36.	Polarity Device	83.	Automatic Selection Control or Transfer Relay
37.	Undercurrent or Underpower Relay	84.	Operating Mechanism
38.	Bearing Protective Device	85.	Carrier of Pilot Wire Receiver Relay
39.	Field-Reducing Contactor	86.	Lockout Relay
40.	Field Relay	87.	Differential Protective Relay
41.	Field Circuit Breaker	88.	Auxiliary Motor or Motor Generator
42.	Running Circuit Breaker – Contactor or Switch	89.	Line Switch
43.	Manual Transfer or Selector Device	90.	Regulating Device
44.	Unit Sequence Starting Relay	91.	DC Voltage Directional Relay
45.	DC Overvoltage Relay	92.	DC Voltage and Current Directional Relay
46.	Reverse-Phase or Phase-Balance Current Delay	93.	Field Changing Contactor
47.	Phase-Sequence Voltage Relay	94.	Tripping or Trip-Free Relay

IEEE SUFFIX LETTERS

Auxiliary Devices

Suffix	Device	Suffix	Device	Suffix	Device
X	Auxiliary Relay	O	Opening Relay	OP	"b" Auxiliary-Switch Relay
Y	Auxiliary Relay	C	Closing Relay	U	"up" Position Relay
Z	Auxiliary Relay	CS	Control Switch	D	"down" Position Relay
R	Raising Relay	CL	"a" Auxiliary-Switch Relay	PB	Pushbutton
L	Lowering Relay				

Valves

Suffix	Device	Suffix	Device	Suffix	Device
BD	Blowdown	PR	Pressurizing	C	Closing Relay
BP	Bypass	RL	Recycling	CS	Control Switch
CH	Check	RF	Relief	CL	"a" Auxiliary-Switch Relay
D	Discharge	S	Suction	OP	"b" Auxiliary-Switch Relay
EP	Electro-Pneumatic Instrumentation	TT	Throttling Control	U	"up" Position Relay
		VT	Vent	D	"down" Position Relay
IS	Isolation	L	Lowering Relay	PB	Pushbutton
MA	Manifold	O	Opening Relay		

INDUSTRIAL SYMBOLS

Device	Symbol	Device	Symbol
FUSE		CIRCUIT BREAKER (THREE POLE)	
CURRENT TRANSFORMER		COIL (AIR CORE)	
RELAY, TRANSFER CONTACTS		COIL (MAGNETIC CORE)	
		COIL (TAPPED)	
		COIL (ADJUSTABLE)	
CIRCUIT TERMINAL			
TERMINAL BOARD OR STRIP		TRANSFORMER (AIR CORE)	
FEMALE CONNECTOR			
MALE CONNECTOR		TRANSFORMER (MAGNETIC CORE)	
ENGAGED CONNECTORS		AUTOTRANSFORMER	
CIRCUIT BREAKER (SINGLE POLE)			

TRANSFORMER CONNECTIONS. . .

1ϕ

3ϕ, GROUP 1, 0° ANGULAR DISPLACEMENT

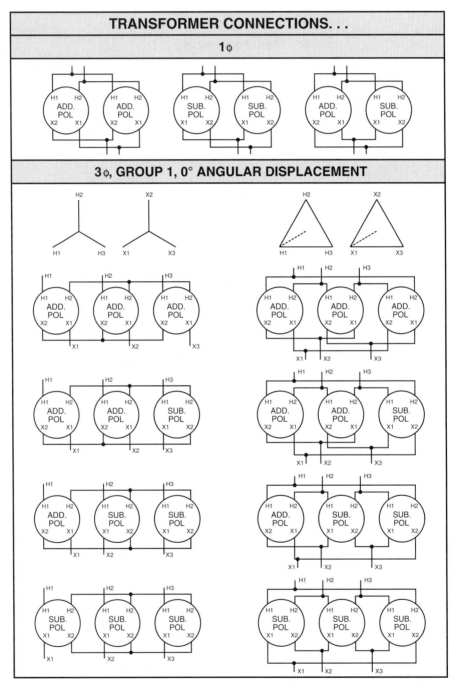

continued. . .

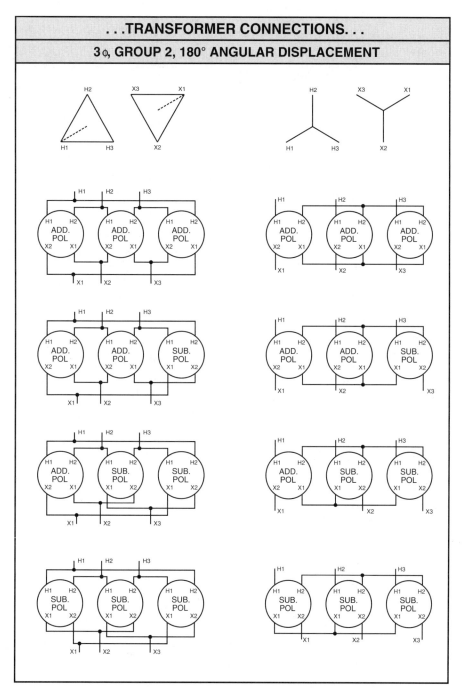

...TRANSFORMER CONNECTIONS...
3 φ, GROUP 2, 180° ANGULAR DISPLACEMENT

continued...

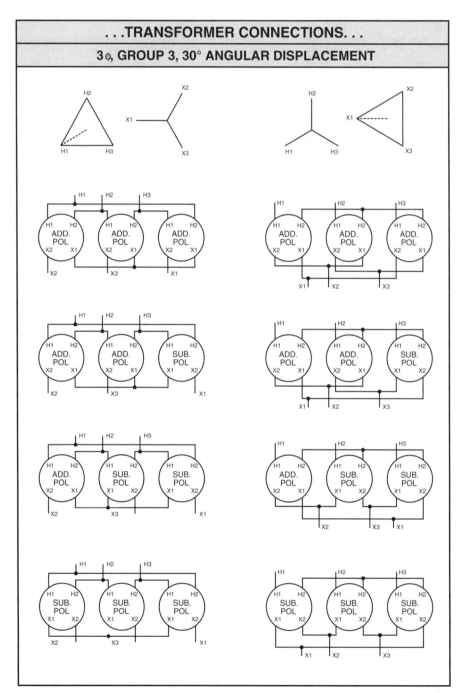

. . .TRANSFORMER CONNECTIONS. . .

3 φ, GROUP 3, 30° ANGULAR DISPLACEMENT

continued. . .

. . .TRANSFORMER CONNECTIONS. . .

3φ TO 6φ, GROUP 4, 0° ANGULAR DISPLACEMENT

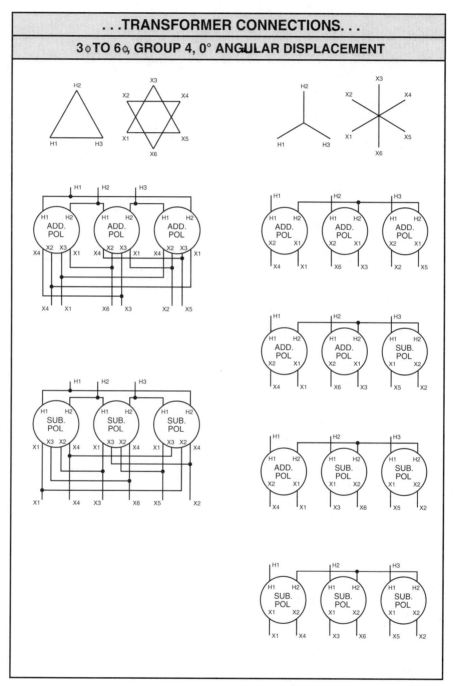

continued. . .

. . .TRANSFORMER CONNECTIONS

3 φ TO 6 φ, GROUP 5, 30° ANGULAR DISPLACEMENT

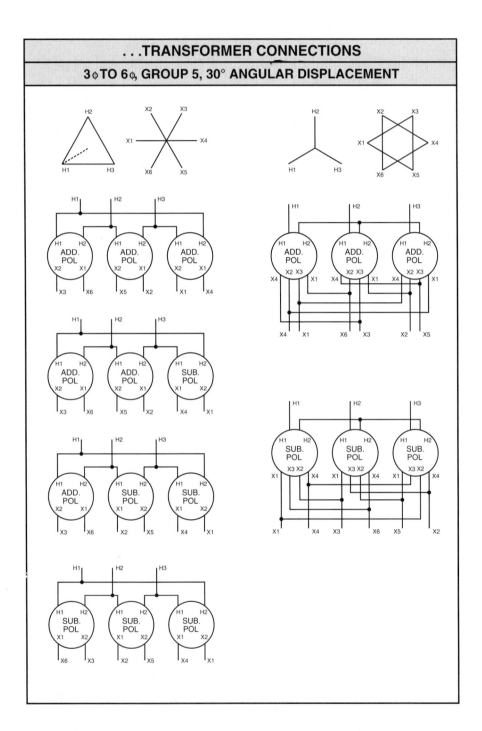

GLOSSARY

A

air core: Coils of transformers which have no iron in their magnetic circuit, used chiefly in radio frequency circuits.

alnico: An alloy of iron with aluminum, nickel, and cobalt.

alternating current (AC): Current that reverses its direction of flow twice per cycle.

ambient temperature: The temperature of the air around a piece of equipment.

ampacity: The amount of current for which a conductor or device is rated.

ampere (A): The current flowing through 1 Ω resistance at 1 V of electrical pressure. The movement of 6.28×10^{18} electrons, at a uniform rate, past a given point in a circuit in one second equals 1 A.

ampere-turns: A unit of magnetomotive force which is obtained by multiplying the current in amperes by the number of turns in a coil.

angle of lag or lead: Phase angle by which quantities, such as voltage, current, or impedance may follow or precede one another.

apparent power: The power of an alternating current circuit where the product of the effective voltage and current values is expressed in terms of volt-amperes (VA). The practical unit of apparent power is kilovolt-ampere (kVA) which is 1000 VA.

arc: An intense luminous discharge formed by the passage of electric current across a space between electrodes or conductors. The discharge is caused by ionization of the air, vapor, or gas between the electrodes.

askarel: A synthetic, nonflammable, insulating liquid which, when decomposed by the electric arc, involves only nonflammable gaseous mixtures.

autotransformer: A transformer in which at least two windings have a common section.

B

back electromagnetic force (counter electromotive force): The voltage created in an inductive circuit by a changing current flowing through the circuit. The polarity of the back electromagnetic force is opposite to the applied voltage, but is never greater in amplitude than the applied voltage.

balance coil: A single-winding transformer having a tap located at some point between its two ends.

bar current transformer: A window current transformer with a solid bar placed permanently through the window.

burden: Admittance of the secondary circuit expressed in siemens and power factor (lagging or leading). The load (burden) placed on instrument transformers is normally expressed as the apparent power in VA absorbed at a specified power factor and at the rated secondary voltage.

bus: An electrical conductor which serves as a common connection for two or more electrical circuits. A bus may be in the form of rigid bars, either circular or rectangular in cross section, or in the form of stranded conductor overhead cables held under tension.

C

cable: One or more insulated or non-insulated wires used to conduct electrical current or impulses.

capacitance: The ability to store energy in the form of an electrical discharge.

capacitor: An electronic device used to store an electrical charge.

capacity: The load for which a piece of electrical apparatus is rated. The capacity is given on manufacturer's nameplate.

choke coil: An inductor used to limit the flow of alternating current.

circuit: A conductor, or a system of conductors, through which an electric current flows or is intended to flow.

closed-core transformer: A transformer having a magnetic circuit consisting completely of iron.

coil: A number of turns of wire wound on an iron core on a coil form made of insulating material. A coil offers considerable opposition to the passage of AC, but very little opposition to DC.

conductance: The measure of the ability of a component to conduct electricity. Conductance is expressed in mohs.

constant-current source: A voltage source in which the voltage impressed on the secondary series circuit is automatically increased or decreased as the current decreases or increases, keeping the current of the secondary circuit at a constant value.

constant-current transformer: A transformer which furnishes current at a constant amperage to a load of prescribed limitations.

constant-voltage transformer: A transformer whose output voltage is maintained at a constant.

corona effect: An electrical discharge which appears as a glow of colored light on the surface of a conductor when the potential exceeds a certain critical value. The corona effect is due to ionization of gases by high voltage. See *St. Elmo's fire.*

copper loss: The energy loss in the windings when a transformer is loaded.

counter electromagnetic force: See *back electromagnetic force.*

coupling: The means by which electrical energy is transferred from one circuit to another. Coupling can be direct through a conductor, electrostatic through a capacitor, or inductive through a transformer.

current: The amount of electrons flowing in a circuit. Current is measured in amperes, milliamperes, and microamperes.

current-carrying capacity: The maximum current that can be continuously carried without causing a permanent change in electrical or mechanical properties of a device or conductor.

current-limiting reactor: A reactor which limits the current that can flow in a circuit under short-circuit conditions or under other operating conditions, such as starting, synchronizing, etc.

current transformer (series transformer): A transformer that has the primary, normally consisting of one or more turns, connected in series with the line.

customer: An individual, firm, or organization which purchases electrical service at one location under one rate classification, contract, or schedule.

cycle: One complete wave of alternating voltage or current.

D

decibel (db): The unit used to measure the intensity level of sound.

delta connection: Connection in which the windings of a 3φ transformer are connected in series to develop a closed circuit. The connection actually forms a triangle like the Greek letter Δ (delta).

dehydrating breather: A device containing calcium chloride which is connected to a pipe entering the tank above the oil level.

demand: The rate at which electrical energy is delivered by a system, part of a system, or a piece of equipment. Demand is expressed in kilowatts (kW).

dielectric: The insulating material between the plates of a capacitor or adjacent wires in a cable.

direct current (DC): Current that flows in one direction only.

distribution: The act or process of distributing electric energy from convenient points on the transmission or bulk power system to the consumers.

distribution transformer: A transformer that changes the primary feeder voltage to the voltage required by the consumer.

E

eddy current: Unwanted current induced in the metal field structure of the motor, due to the rate of change in the induced magnetic flux.

effective current: The value of AC which causes the same heating effect as a given value of DC. For sine wave alternating currents, the effective current is 0.7071.

efficiency: The ratio of output to input.

electromagnetic field: A magnetic field located at right angles both to the magnetic flux lines and to their direction of motion.

energize: To apply voltage required for operation of any equipment.

energy: The capacity to do work.

excitation: The power required to energize the magnetic field of transformers, generators, and motors.

exciting current: The current which the magnetic circuit draws from the line when it is excited at normal voltage and frequency with no load on the secondary.

explosion vent: A pipe, 4″ in diameter or greater, which extends a few feet above the cover of a transformer and is curved toward the ground at the outlet end of the pipe.

F

film cutout: A device in the base of a lamp that punctures and re-establishes the circuit when subjected to a voltage above a certain critical value.

flux density: The measure of the magnetic flux lines per unit area taken at right angles to the direction of flux.

frequency: The number of complete electrical cycles per second.

fuse: An overcurrent protection device consisting of a piece of wire which melts and breaks when the current through it exceeds the rated value of the fuse.

G

generator step-up transformer (power transmission transformer): A transformer that steps up the generator voltage to a high voltage for economical power transmission purposes.

gilbert: A unit of measurement based on one oersted-centimeter.

ground: A connection, intentional or accidental, between an electric circuit and the earth or some conducting body or chassis serving in place of the earth.

grounded: Connected to earth or to some conducting body that serves in place of the earth.

ground fault: A condition in which current from a hot power line is flowing to the ground.

ground wire: A conductor leading from an electrical device to an electrical connection with earth.

H

henry: The inductance in which a current changing its rate of flow 1 A per second induces an electromotive force of 1 V.

hertz (Hz): The international unit of frequency equal to one cycle per second.

hysteresis: The magnetic flux lines that lag behind the magnetizing force that causes them.

I

impedance: The opposition to alternating current.

impedance voltage: The voltage impressed on the transformer to circulate normal full-load current.

induced voltage: A voltage produced in a circuit by changes in the number of magnetic flux lines which are linking or cutting across the conductors of the circuit.

inductance: The property of an electrical circuit by which a varying current induces an electromotive force in the circuit or in a neighboring circuit.

inductive coupling: A form of coupling in which energy is transferred from a coil in one circuit to a coil in another circuit by induction.

instantaneous voltage value: The amplitude of voltage in an AC wave at any selected instant in the cycle.

instrument transformer: A transformer that steps down the voltage or current of a circuit to a low value that can be effectively and safely used for the operation of instruments, such as ammeters, voltmeters, wattmeters, relays used for various protective purposes, telemetering used for indications at remote areas, and dispatching energy.

insulation: Any non-conductive material used to prevent the leakage of electrons from a conductor.

insulator: A device having high electric resistance, used for supporting or separating conductors to prevent an undesired flow of current between conductors or to other objects.

ionization: The breaking up of a gas atom into two parts, a free electron and a positively charged ion.

isolated neutral system: A system with no intentional connection to earth except through indicating, measuring, or protective devices of very high impedance.

K

kilovar (kVAR): 1000 reactive VA.

kilovolt (kV): 1000 V.

kilowatt (kW): 1000 W.

kilowatt-hour (kWh): The basic unit of electric energy equal to 1 kW of power supplied to or taken from an electric circuit steadily for one hour.

L

lagging load: An inductive load in which the current lags behind the voltage.

latching relay: A relay which employs a mechanical device to latch and hold the contacts in position after the actuating coil is de-energized.

leading load: A capacitive load in which the current leads the voltage.

leakage current: Undesirable flow of current through or over the surface of an insulating material or insulator.

leakage flux: Magnetic flux lines that do not encircle all the turns in a coil or transformer, thereby not contributing to inductance or the transfer of energy from one coil to another.

leakage inductance: A self-inductance caused by leakage flux generated in the winding of a transformer.

leakage lines: The magnetic flux lines produced by the primary coil which do not pass through the secondary. The greater the number of leakage lines, the less the useful effect realized.

leakage reactance: The result of leakage in a transformer.

left-hand rule: The relationship between the current in a conductor and the magnetic field existing around the conductor.

lightning arrester: A protective device which prevents lightning damage.

line voltage regulator: A device, such as a special transformer, that delivers an essentially constant voltage to the load.

load: The amount of electric power delivered or required at any specified point or points in a system. The load originates primarily at the power-consuming equipment of the customer.

loss: Energy dissipated without accomplishing useful work.

M

magnetic induction: The setting up of magnetic flux lines in a material by an electric current.

magnetic leakage: The portion of magnetic flux that passes through either one of the coils but not through both.

magnetic permeability: The measure of the ease with which certain substances conduct magnetic flux lines.

magnetomotive force (MMF): The magnetic pressure that determines the number of magnetic flux lines through a medium.

maxwell: The unit of measurement of the total number of magnetic flux lines in a magnetic field.

mutual coupling: Coils located in such a way that a varying current in one coil induced a voltage in an adjacent coil.

mutual inductance: The flux linkage between two coils.

N

nameplate rating: The full, continuous rating of a piece of electrical equipment under specified conditions designated by the manufacturer.

network: A system of transmission or distribution lines which are connected and operated to permit a multiple power supply to any major point.

O

oersted: The magnetomotive force per unit length. The magnetizing force is found by dividing the magnetomotive force (in gilberts) by the length of the magnetic circuit (in cm).

ohm: The basic unit of resistance.

open core transformer: A transformer having a magnetic circuit consisting partly of air and partly of iron.

oscillation: Fluctuations in a system or circuit, especially those consisting of the flow of charges alternately in opposite directions.

out-of-phase: Having waveforms that are of the same frequency but not passing through corresponding values at the same instant.

output: Useful energy delivered.

overload: The condition that occurs when circuit current rises above the normal current level at which the load and/or circuit is designed to operate.

overvoltage taps: Taps used for voltages above the normal line voltage and maintain a constant-rated voltage on the secondary winding.

P

permeability: The measure of the ease with which a given material can carry magnetic flux lines.

permeance: The measure of ease with which magnetic flux permeates a material. Permeance is the reciprocal of reluctance.

phase displacement: The difference in phase between primary voltage and secondary voltage vectors. Phase displacement is positive when secondary voltage vector leads primary voltage vector.

polarity: The particular state of an object, either positive or negative, which refers to the two electrical poles, north and south.

polyphase: A term used to indicate three or multiples of three phases. Polyphase circuits normally involve three alternating voltages spaced 120° apart.

potential transformer: An instrument transformer in which the secondary voltage, in normal conditions of use, is substantially proportional to the primary voltage and differs in phase from it by an angle which is approximately zero for an appropriate direction of the connections.

power: The rate of doing work. Power is expressed in watts.

power factor: The ratio of real power (kW) to apparent power (kVA) for any given load and period of time.

power reactive: The portion of apparent power that does no work. It is measured in kVAR.

primary voltage: The voltage of the circuit supplying power to a transformer.

primary winding: The coil which is energized from a source of alternating voltage and current.

R

rated frequency: The value of the frequency on which an electrical device is rated.

rated primary voltage: The value of the primary voltage which appears in the designation of the transformer and on which its performance is based.

rated secondary voltage: The value of the secondary voltage which appears in the designation of the transformer and on which its performance is based.

rated transformation ratio: The ratio of the rated primary voltage to the rated secondary voltage.

reactance: The measure of the induced voltage due to the inductance of a circuit. Reactance is denoted by the letter X and is measured in ohms.

reactive current: Current out-of-phase with the voltage by 90°. Reactive current does not represent real energy.

reactor: A device whose primary purpose is to introduce reactance into a circuit.

real power: The work-producing part of apparent power. It is the rate of supply of energy and is measured in kW.

regulation: The difference in voltage between no load and full load expressed as a percentage of the full-load voltage.

relay: An electromagnetic switch employing an armature to open or close contactors.

reluctance: A measure of the force exerted against magnetic flux which permeates a material.

remanence: The magnetic induction remaining in a magnetic circuit after removal of applied magnetomotive force.

resistance: The non-reactive opposition which a device or material offers to the flow of current. It is measured in ohms and is normally denoted by the letter R.

retentivity: The ability of a substance to retain magnetic flux lines after the magnetizing force has been removed.

right-hand rule: The relationship between the movement of an inductor and the induced voltage in the inductor as it cuts magnetic flux lines.

right-hand screw rule: The direction of advance of a right-hand screw corresponds with the direction of current flow through a conductor, while the direction in which the screw is rotated corresponds with the direction of the magnetic flux surrounding the conductor.

rotary transformer: A rotary converter.

S

sampling valve: A needle valve which is used to draw a sample of liquid for test purposes.

saturation point: The point where an increase in magnetizing force (ampere-turns) does not greatly increase the flux density.

saturable-core reactor: A magnetic-core reactor in which the reactance is controlled by changing the saturation of the core through variation of a superimposed flux.

secondary voltage: The output or load supply voltage of a transformer.

secondary winding: The transformer coil which delivers the alternating current and voltage to the load.

series transformer: See *current transformer*.

short circuit: A low resistance connection across a voltage source or between the sides of a circuit or line, normally accidental and normally resulting in excessive current flow which often causes damage.

shunt reactor: A reactor that is connected in shunt to an electrical system. A shunt reactor is used to draw inductive current.

sinusoidal: Varying in proportion to the sine of an angle of time function. Ordinary AC is sinusoidal.

sparkover: Flashover or breakdown of the air between the electrodes permitting the passage of a spark.

spike: A short duration transient.

standard temperature and pressure: A temperature of 0°C and a pressure of 760 mm of mercury.

St. Elmo's fire: Visible corona-like display of electric discharge sometimes seen around a metal point. See *corona effect*.

step-down transformer: A transformer in which the primary winding has a greater number of turns than the secondary winding.

step-up transformer: A transformer in which the secondary winding has a greater number of turns than the primary winding.

substation: An assemblage of equipment installed for switching, changing, or regulating the voltage of electricity.

surge: A sudden and transient variation in the current and/or potential at a point in a circuit, such as that due to sudden application of a voltage.

switching station: An assemblage of equipment for the sole purpose of tieing together two or more electric circuits or systems.

system voltage: The rms line-to-line voltage measured at a given point in an electrical power system.

T

tap: A connection brought out of a winding at some point between its limits to allow changing the voltage or current ratio.

tap changer: A selector switch device used to change transformer taps with the transformer de-energized.

teaser transformer: A transformer connected between the midpoint of the main transformer and the third-phase wire of the 3ϕ system.

tertiary winding: A third winding that is often used in power transformers to provide station power requirements or a tie with synchronous condensers.

transformer: A device which steps-up or steps-down alternating voltages.

transient: A temporary component of current existing in a circuit during adjustment to a changed load, different source voltage, or line surge.

transmission: The act or process of transporting electric energy in bulk from a source or sources of supply to other major parts of the system or to other utility systems.

transmission line: A line used for bulk transmission of electrical energy.

turns ratio: The ratio between the voltage and the number of turns on the primary and secondary windings of a transformer.

U

unity coupling: Perfect magnetic coupling between two coils so that all the magnetic flux produced by the primary winding passes through the entire secondary winding.

unity power factor: A power factor of 1.0 (100%) obtained when current and voltage are in phase as in a circuit containing only resistance or in a reactive circuit at resonance.

V

variable transformer: A continuously adjustable autotransformer consisting of a single layer of wire wound on a toroidal core and a carbon brush which traverses this winding.

vector: A quantity involving direction and magnitude. The length of the vector represents the magnitude. The direction of the vector indicates the direction or angle of the quantity from a given reference axis.

volt (V): The basic unit of electrical pressure. It is measured in ohms.

voltage drop (VD): The voltage developed between the terminals of a circuit component by the flow of current through the resistance or impedance of that part.

W

watt (W): The electrical unit of power or rate of doing work. One HP is equivalent to 746 W.

winding: One or more turns of wire forming a continuous coil.

window current transformer: A transformer that consists of a cylindrical-ring core of thin iron lamination similar to a stack of washers.

wound-current transformer: A transformer that has separate primary and secondary windings mounted on a laminated core.

wye connection: A connection in which one end of each of the windings of a polyphase transformer is connected to a common point (the neutral point) and the other end is connected to its appropriate line terminal.

wye-wound: A method of winding coils for use in 3ϕ current circuits. The coils have one common connection and their ends extend outward to form the letter Y.

INDEX